Which Way Is the Sky Falling?

Thomas G. Kyle

NorthWord
PRESS, INC

MINOCQUA, WISCONSIN

Edited by Greg Linder and Kim Donehower
Designed by Patricia Bickner Linder

Published by: NorthWord Press, Inc.
P.O. Box 1360
Minocqua, Wisconsin 54548

For a free catalog describing NorthWord's line of nature books and gifts, call 1-800-336-5666.

Library of Congress Cataloging-in-Publication Data

Kyle, Thomas G. (Thomas Gail), 1936-
    Which way is the sky falling? / by Thomas G. Kyle.
       p. cm.
    ISBN 1-55971-216-3
  1. Environmental degradation. I. Title.
GE140.K94 1993
363.7--dc20   93-19628
                                 CIP

Printed in U.S.A.

TABLE OF CONTENTS

Preface

Concern about the environment inspired this book. However, this is something other than a book announcing that "the sky is falling." Scaring people about environmental problems is easy. Instead of generating fear, I hope to generate understanding, and to bring out problems that television and the press often ignore.

Although environmental abuse threatens our civilization, many environmental fears that are spotlighted by the media are not too important. They and some scientists take advantage of our fears because we lack understanding. Hazards that don't photograph well, or that have little prospect of producing research funds for scientists, may get ignored.

The solution is a better-informed public—people who know how the environment works, not just viewers programmed to shudder at the mention of global warming. People who love the environment want to understand how it all fits together, but don't need a graduate course in complex details. They need a briefing on what scientists do and do not know about topics like global warming and ozone holes. These are the things I address. Right now, people worry about future temperatures and skin cancer without knowing much about them. If they knew more, they might fear less—or at least have a more realistic idea of what faces them in the future.

Too many people feel intimidated by scientific knowledge. They should not. A common-sense understanding of "what affects what" is not much different from what the best scientists can predict. Scientists have theories and rooms full of data, but when it comes to predictions of the future, their crystal ball is filled with smoke. This book lays out what is known, describes the limits to our knowledge, and clearly explains the causes and effects of global warming.

No one ever acknowledges that scientists can predict almost nothing with confidence. Inability to predict means that giant hazards may get overlooked, because they have not been foreseen. Therefore, changing our ways when a new hazard is recognized may mean we're changing too late.

Consider the ozone hole. Even if we never release another molecule of ozone-destroying chemicals, the hole will not heal for 50 or 100 years. Read on, and find out why.

1

What Survives if the Ozone Goes?

Keeping a good tan will be easy.

Because there has been so much alarm spread about the ozone hole by the media, a few calming words are necessary before beginning discussion of the ozone hole. The calming words are nothing more than a realistic review of the ozone situation—providing the facts instead of the one-liner headlines people usually hear.

One survey found that 96 percent of adults know about the ozone hole. Surveys usually find that a quarter of the population cannot name the current president, and even fewer can locate Africa on a world map. The 96 percent figure seems unbelievable, but there is no denying that the ozone hole has been a regular feature in the news. The media has, however, neglected to give the public the full story.

The loss of ozone is a serious problem. Saying that the media mislead people is not saying we should keep pumping out ozone-eating chemicals like chlorofluoro-carbons (CFCs). We should, though, understand why the chemical releases have to stop.

A few people contend that something other than CFCs caused the ozone hole. They have yet to prove their case. Until they do, getting rid of ozone-destroying chemicals is the only prudent action to take. The hazards we face from reduction of ozone may have been overstated by some scientists, but the loss of ozone has to stop.

The ozone hole is not a year-round feature. The hole lasts for two months during the Antarctic spring; then ozone levels return to something not so different from historic levels. During the long Antarctic winter there is no sun, but chemical events take place in the dark, frigid, upper atmosphere. These events set the stage for the "return of the hole."

Come spring, sunlight drives photochemical reactions that quickly destroy ozone. However, the sun also warms the upper atmosphere, which actually upsets the conditions necessary for ozone destruction. After a couple of months, the upper atmosphere has warmed enough to stop ozone destruction; in fact, the sunlight manufactures new ozone.

When the ozone thins, solar ultraviolet light reaches the ice and ocean below. In spite of the fact that 70 percent of the ozone is gone at times, not all that much ultraviolet light gets through. The polar regions contain more ozone than the lower latitudes, so some protection remains. One scientific journal article explained the ultraviolet leaking through the ozone hole in this way: "The amount of ultraviolet getting through was comparable to the amount received on a summer day near Chicago." This may be enough to damage some sun-sensitive Antarctic creatures, but not most.

According to some accounts, three percent of the ozone may have disappeared from the protective layer above the inhabited United States. Here's an intriguing thought: If everyone moved 60 miles north, the move would compensate for a three percent loss of ozone.

OZONE AND CANCER

Cancer seems to be the most negative word in our language, and we often hear that loss of ozone could cause many thousands of cases of skin cancer. When putting this in perspective, it is important to understand that most skin cancers are not malignant, or life-threatening. Mentioning that ultraviolet light causes skin cancers induces a touch of terror, but you might even compare some skin cancers to warts. No one wants warts, even though they pose little threat to long-term health. A similar procedure gets rid of both warts and some skin cancers. A ten-minute visit to a dermatologist's office is all it takes to remove them. Many times, the removal site does not even need a band-aid.

But don't be entirely reassured. Malignant melanoma is a skin cancer that is very serious indeed. It does kill. Some other skin cancers can also be malignant. However, not many fall into this category.

OZONE AND THE MEDIA

It is shameful when a problem gets exaggerated until discussion reaches the level of hysteria. Once the public alarm is set off, the public may demand actions that are unnecessary, cost too much, and might even be harmful in the long run. Such a response is driven by fear, and fear of the ozone hole has been fostered by television, magazines, and politicians. Before he became vice president, then-Senator Gore made statements like, "We have to tell our children that they must redefine their relationship to the sky," and, "They must begin to think of the sky as a threatening part of their environment." It would be tragic indeed if children began to see the sky as a threat looming above them.

Time magazine once showcased an ozone hole story with a front cover headline saying, "Vanishing Ozone: The Danger Moves Closer to Home." Inside, the story claimed, "This unprecedented assault on the planet's life-support system could have horrendous long-term effects on human health, animal life, the plants that support the food chain, and just about every other strand that makes up the delicate web of nature."

Michael Lemonick, who wrote the *Time* story, later rationalized such strong language in this way: "I don't think we say in the article that the reader is going to 'get fried.' We say there is strong evidence of the possibility." Lemonick admits that the press tends to play up the more dramatic aspects of the ozone situation and play down the less dramatic concerns. He also says, "I think it's true to some degree that the public cannot deal with the subtleties of complicated issues."

Charles Alexander, the senior editor

who coordinates *Time's* science coverage, told a Smithsonian Institution conference on environmentalism: "On this issue, we have crossed the boundary from news reporting to advocacy."

Dianne Dumanoski of the *Boston Globe* told the same conference: "There is no such thing as objective reporting. I've become even more crafty about finding the voices to say the things I think are true."

As the result of such "reporting," the public is presented with the opinions of non-technical reporters who seldom understand the problem—much less the solution. Reporters reach their own conclusions about subjects that even experts do not agree upon. Having decided what is true or right, the media presents its misinformation to the public.

The nonpartisan Center for Science, Technology, and Media hired the Gallup organization to survey 400 climate experts about global warming. The Center then analyzed the content of broadcasts by nine major media outlets to see if the media represented the experts' consensus. It did not. The Center's conclusion was, "The public is receiving a skewed portrait of the scientific debate over global warming. In news accounts, scientists are presented as relatively unified supporters of global warming theory. Yet the Gallup survey revealed considerable debate and uncertainty about global warming theory within the scientific community."

All of this aside, the loss of ozone is a worrisome problem. Dangerous events will occur as ozone disappears—things like an increased number of deaths from malignant melanoma and decreased crop productivity. The loss of ozone may well lead to serious problems that haven't yet been anticipated. These prospects are cause for worry, but panic is not productive. Any time human activities alter nature, we must carefully examine the situation—whether the alteration is the accidental destruction of ozone or the intentional elimination of mosquitoes.

Unfortunately, a symbiotic relationship exists between reporters and some scientists. Reporters want dramatic stories. The scientists want to ensure that money is available for their research. Reporters call the scientists, knowing that the scientists acquired research funds by telling a dramatic story. "Nothing to worry about" is never a dramatic story. So the reporter gets a pessimistic report, and the scientist gets nationwide coverage. The coverage makes the scientist's next grant easier to obtain. The scientist's name becomes well known, his institution gets publicity, and the magazine sells more subscriptions. Everybody gains, nobody loses—except the poorly informed public.

Of course, not all of the scientists play this game. Melvyn Shapiro, Chief of Meteorological Research at the National Oceanic and Atmospheric Administration Laboratory (NOAA) in Boulder, Colorado, is one who does not. He has spent 20 years researching ozone and speaks with authority. Here is what he says about science and the media relating to the ozone problem:

"If you have a doomsday scenario, you get a lot of money. Research organizations are in great competition with each other to get the politicians' ears and obtain the necessary resources. If you want money, you have to come up with a doomsday scenario.

"The public doesn't understand how speculative these scenarios are. And if a member of the lay public reads an article such as the *Time* cover story on ozone, he or she gets the impression of imminent disaster—the ozone is being destroyed! This is imminent danger! The lay reader cannot interpret the few qualifiers—'if,' 'could,' 'might,' 'possibly'—studded through the article. The reader just thinks he's going to get fried."

Shapiro offers a frightening scenario of his own, if misinformation prevails in handling the ozone destruction question:

"What people are going to have to deal with in five years is that they'll be told they have to buy a new refrigerator because 'We can't recharge your old one.' "

The air conditioner that breaks down during the heat wave will meet a similar fate, according to Shapiro: "Sorry, [it] won't accept the new chemistry. You'll have to come up with $1,500 for a new one."

One example of the way in which premature scientific findings can become a national scare involves a press conference held by NASA on February 3, 1992. The event was so obviously a manipulation of the media and the public that two magazines, *Insight* and *Reason,* ran stories debunking the news conference and its information.

At the press conference James Anderson, a Harvard chemist, described a flight undertaken by a converted ER-2 spy plane into the Arctic vortex. The flight had encountered a higher level of CFCs than expected, and this finding was used to project an ozone hole at the North Pole that would extend into populated areas. This prediction was couched in the term "increasingly likely."

The flight in question had involved work performed by 40 scientists. Some later said they were unaware that the press conference was to take place. Others said they were muzzled, cautioned not to speak to reporters. The scientists were from NASA, NOAA, and NCAR (National Center for Atmospheric Research). One of the scientists involved in the program insisted on anonymity and said, "NASA is the 800-pound gorilla in the ring. You either go along with the gorilla or you stay out of its way."

NASA held the press conference two months before the end of its project. When the predicted hole failed to appear and satellite measurements of chlorine oxide (the chemical that destroys ozone) were found to have diminished, some members of the scientific team wanted to issue a second interim report. NASA nixed the idea. According to *Reason,* the press conference had been timed to bolster the agency's request for its global climate change program, whose funding was slated to double in 1993. This is one example of media manipulation. Other examples have received less coverage because they were not so blatant. Underlying all such manipulations, though, is the often-unmentioned threat of ultraviolet radiation reaching the surface of the earth.

ULTRAVIOLET AND US

Ultraviolet is nothing but light. When early researchers discovered light whose wavelength was too short for the human eye to see, they called it ultraviolet. The shortest wavelengths we can see are violet.

Ultraviolet is nothing special in terms of light. We cannot see it, but that is because our eyes have shortcomings. In the area of special effects, though, ultraviolet is different. It can make things glow in the dark, or fluoresce. It does so by kicking molecules into excited states. Visible light is too weak to provide such a strong kick. The kick depends on the light's wavelength—i.e., frequency—not the light's brightness or intensity.

The ultraviolet getting through holes in the ozone will not cause us to glow in the dark (although ultraviolet can make some false teeth glow). However, the extra kick that ultraviolet carries can break molecules apart or promote unusual chemical reactions.

Ironically, nature forms ozone in the first place by using ultraviolet's ability to tear up molecules. Ultraviolet from the sun tears apart an oxygen molecule consisting of two oxygen atoms. One oxygen atom then joins up with an undisturbed oxygen molecule to form a new molecule containing three oxygen atoms. The newly joined trio is an ozone molecule. Ultraviolet also breaks up other molecules, like CFCs. The breakup of CFCs releases chlorine that, with the help of more ultraviolet, can destroy ozone.

Closer to earth, ultraviolet can harm you. Ultraviolet affects your largest organ—your skin. Nearly everyone has already suffered damage from ultraviolet. The damage is called sunburn. The extra "kick" in the rays breaks chemical bonds within the skin, leaving it red and sore.

Two different bands of ultraviolet often get mentioned: UV-A, and UV-B. Figure 1 shows the wavelength interval of both. The bands are indicated under a curve showing how much ultraviolet radiation comes from the sun. The curve is the amount

reaching the ground. There is not much UV-B to start with, and the ozone takes care of most of it. Enough gets through the ozone layer, however, to cause sunburn.

Another different feature of ultraviolet light is the direction from which it comes. The sky is blue because air scatters more short wavelength radiation. The blue that lights up the sky is really sunlight that was headed somewhere else before a molecule of air scattered it. And if the sky is blue because air scatters more blue light than red or yellow, imagine how much more ultraviolet gets scattered. Ultraviolet is even shorter in wavelength than blue light, and it scatters more readily.

Air scatters so much ultraviolet that photo-gray sunglasses (the ones that turn dark in bright light) sometimes turn dark even when placed inside an automobile where no sunlight reaches them. This happens because ultraviolet light—not bright light—darkens photo-gray glasses. Enough

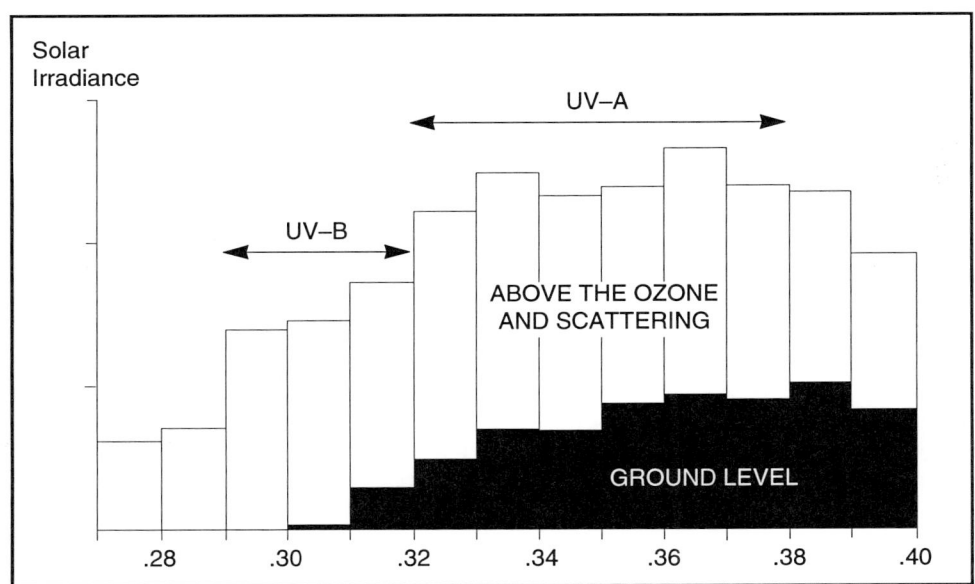

FIGURE 1 *The curve represents the amount of ultraviolet radiation above the atmosphere and at ground level. Most of the losses in the 0.36 to 0.40 micrometer wavelengths is due to scattering by air molecules other than ozone. UV-B is the light that causes skin to tan and causes skin cancers.*

ultraviolet scatters into the car to darken the glasses.

Clouds completely block the long wavelength infrared radiation because water is a strong absorber of infrared. Water only absorbs small amounts of UV-B radiation, however—ultraviolet can penetrate several feet into the ocean. This explains why people can get sunburned on cloudy days. Water absorbs little ultraviolet, so its radiation passes right through the clouds. Of course, not all of it gets through. Much of it gets scattered. And so trying to get a suntan on a cloudy day generally yields disappointing results.

Sunburn is not the only ultraviolet-induced problem. Skin cancer is a delayed reaction to sunburn, delayed by as much as 20 or 30 years. It's commonly accepted that ultraviolet can also cause other health problems. Most of the causal relationships remain unverified, but they seem to make sense, and ultraviolet probably does contribute at least a little. These problems include cataracts and suppression of the immune system.

Cataracts cause loss of vision by forming a film on the eye. New laser surgery makes removal of the film an outpatient operation, but not a trivial one like removing simple skin cancers. Until lasers modernized cataract surgery, it was a difficult experience for the patient.

There have long been reports that fluorescent lights contribute to cataracts because they emit some ultraviolet. Ultraviolet light cannot get into the interior of the eye. The cornea stops some of it, but the lens, where cataracts occur, gets the big dose. Strong doses of ultraviolet cause chemical changes in the protein that forms the lens.

It is also possible for the cornea to get what might be termed a sunburn. Such damage, called snow blindness, is common when people spend a lot of time around sunlit water, sand, or snow. Their eyes water and hurt for a few days before they recover. There are some statistical correlations between fluorescent lights and cataracts, but nothing definitive enough to spur regulations.

Cataracts usually affect older people, perhaps because it takes many years of cumulative damage for the cataract to develop; or perhaps because the cataracts develop years after a person is exposed as a youth. However, age is not the only factor. More cataracts occur at lower latitudes, for example. People at such latitudes are exposed to more ultraviolet because they are protected by less ozone. Again, the polar regions have always had considerably more ozone than the lower latitudes.

Stories about suppression of the immune system by ultraviolet come from experiments that may be legitimate, but whose results get stretched beyond all reason. For example, scientists who transferred cancer between animals found that more "successful" transfers occurred after exposing the animals to ultraviolet light. "Successful" in this case meant that the transplanted tumors continued to grow. It takes a good imagination to claim these results indicate that ultraviolet promotes herpes and accelerates AIDS. But someone in the scientific community had the imagination.

The problem with proving that ultraviolet exposure suppresses the immune system is stress. Get a sunburn, feel pain, and stress builds up. Expose animals or people to the amount of ultraviolet needed to conduct tests, and they get burned and stressed. Stress is a well-known suppressor of the immune system. Thus, it is difficult to know whether the ultraviolet itself suppresses the immune system or whether the stress resulting from exposure is the culprit. All of our worry about the ozone hole causes stress, for that matter, so ozone hole stories may also affect your immune system.

Not all ultraviolet effects are bad. Exposure causes the skin to manufacture vitamin D3, a fat-soluble vitamin that protects against the anomalous bone conditions of rickets in children and osteomalacia in adults. Some say Scandinavians have such light-colored skin because this pigmentation allows them to absorb more ultraviolet. The short winter days in the northern latitudes where they live mean they have only a short time in which to produce vitamin D. However, this theory doesn't explain why Eskimos, who live even farther north, have darker skins.

Of course, ultraviolet exposure is not the only way to get your vitamin D. It can be acquired by eating food and taken directly into the bloodstream in fats. Without the fat, nothing absorbs and carries vitamin D into the blood. Those who are determined to avoid fats face a dilemma: whether to get vitamin D by exposing themselves to fats or to ultraviolet.

Assuming we become smart enough to avoid going out into the sun too much, ultraviolet can still cause problems. It can affect world food production. Ultraviolet radiation can damage plants, much as it sunburns people. It can interfere with photosynthesis and induce genetic damage in plant cells. Experiments reveal how much of a problem this may cause. We will examine the results of these experiments later, after we discuss skin cancers.

SKIN CANCER

The skin we all want to save is a multilayered construction. On the outside is the epidermis, a nonvascular layer, and underneath is the dermis. Within the dermis, cells called fibroblasts make strands of protein, called elastin and collagen. These support the skin, give it moisture, and keep it from wrinkling. Too much sun diminishes the elastin and collagen in the skin, causing it to wrinkle and age. This is a long-term, "serves you right" process that makes moralists rejoice. Those who keep a tan to look healthy when they are young actually damage their skin's ability to produce elastin and collagen. In a balancing of the books, the well-tanned look good when they are young, but not so good before they get very old.

Suntans have not always been considered attractive. Until the 1930s, most people considered tanned skin unappealing. Looking down their noses at tans might have been class snobbery, because people who worked outdoors—common laborers—were tan. Those who were well-off enough to stay indoors were beautifully pale. Perhaps there was ethnic snobbery at work as well. People from England and Scotland have little pigmentation to begin with, while those from southern Europe have more olive complexions.

Today, most people have to spend their days bent over a counter or a computer, far from sunlight. The snob effect has been reversed. Now a tan announces that its wearer has the leisure time to go skiing, to the beach, or to the tanning salon. It says, "This person has a healthy income and time to get away from work."

Becoming prematurely wrinkled is not the only price one pays for having a tan. Some 10 or 20 years later, the penalty of skin cancer may emerge. According to American Cancer Society estimates, 600,000 skin cancers occurred in 1990. When the cancers start popping up, most people develop more than one, so this indicates that perhaps a quarter of a million people developed skin cancer in 1990. Few of these cancers were malignant. Less than one percent caused deaths, but that is still a high price to pay.

Those who died may have failed to get proper treatment. As one dermatologist said, "It takes real talent to die of skin cancer. You have to go around with a hole in your skin for years." Unfortunately, some people manage to do just that.

In spite of the ease and availability of treatment, more people die of common skin cancer than of the feared malignant melanoma. Death rates from melanoma are much higher, but fewer deaths occur. Melanoma represents only three percent of skin cancers, but accounts for 60 percent of the deaths. Each year, 6,300 people die from melanoma. That is a big number. To put it in perspective, though, it represents about a quarter of the number of people who are killed in auto accidents each year.

Even in the case of melanoma, people die indirectly rather than from the lesions on the skin. Cancerous cells from the melanoma break away, and the bloodstream carries them to internal parts of the body. These breakaway cells start a cancer in the region where they settle. The sooner a melanoma is removed, the smaller the chance that its cells will initiate a cancer elsewhere.

You may note that some of the numbers quoted in this chapter are not entirely consistent. The errors are not large enough to be misleading, but comparing numbers too closely may produce confusion. The estimates of different authorities often vary.

Ten times as many melanomas occur today as in 1930—back when sunbathing first became popular. Some 32,000 cases are expected in 1993. In 1930, one person in 500 developed melanoma. Today, it afflicts one out of every 120 persons. Such numbers come from various sources. According to a second set of numbers, the increase since 1930 has been fourfold, not tenfold. There are ways to rationalize the numbers, but the message is the same: Suntans from years ago are now paying off in skin cancers. Dermatologists now classify tans as injured skin.

There is no getting away from discrimination in modern society, and cancer also discriminates. Women develop more melanomas. The greatest increase has occurred in women under 40. In women aged 25 to 29, melanomas are the leading cause of cancer. Dark-skinned people develop fewer skin cancers, because their pigmentation protects them. Olive-skinned people fare much better than northern Europeans. Whites are 60 times more susceptible to skin cancer than blacks.

The more common skin cancers are nonmelanomic basal cell and squamous cell skin cancers. There are also the less common but malignant basal cell and squamous cell carcinomas. Basal and squamous cells are not strange cells caused by UV-B. Basal and squamous are the names of cells that exist in normal skin. Squamous cells are the most common cells in the epidermis. Not too surprisingly, basal cells occur at the base of the epidermis. New cells originate here to replace the skin that keeps wearing away and flaking off. Under the basal cells are the dermis and the fibroblasts that form elastin and collagen.

Basal cell carcinomas are pale, waxy, pearly bumps or scaly red patches. They rarely spread to other organs, but they can lead to the loss of an ear or an eye. Those who develop them should seek treatment at once. It is even more important to treat squamous cell carcinoma quickly. These cells can spread to other organs. They may appear as rough, thick, scaly patches or bleeding, oozing, open sores. These are the sores always mentioned in skin cancer warnings, sores that do not heal.

These cancers, then, will be some of the fruits of reduced ozone. It sounds bad, but anything referring to cancer sounds bad. Northern locations will probably see a larger decrease in ozone and a corresponding increase in skin cancers. Current melanoma rates are six per 100,000 people in Detroit, 11 per 100,000 in Atlanta, and 27 per 100,000 in Tucson. Comparing melanoma cases in cloudy England and sunny Queensland, Australia, provides spectacular evidence concerning melanoma and the sun. British men have an incidence of only two melanomas per 100,000, compared to 40 per 100,000 in Queensland.

If ozone loss does turn out to be greater in the northern states, skin cancer rates there might increase until the chances of skin cancer in Maine are similar to the chances in Texas. Put that way, it does not sound quite so alarming.

Skin does try to protect itself from ultraviolet damage. It tans. The problem is that tanning takes a while. There is time during the average lunch hour to get sunburned. The same sunshine that burns will also yield a tan, but the tan takes 10 or 12 hours to develop. That may happen too late to prevent the initial sunburn, but the tan will nonetheless help prevent the next burn. Tanning occurs when melanocyte cells in the basal layer produce the melanin that darkens the skin. Once formed, the melanin absorbs part of any incoming ultraviolet. Tanning also thickens the skin, a process that contributes to the aging of skin.

DNA in cell nuclei are the target molecules that ultraviolet damages. Ultraviolet causes two nearby DNA units to form an abnormal bond, called a dimer. The body has enzymes to guard against such defective cell nuclei. These enzymes detect the damaged DNA and clip out that cell, effectively destroying it. Other enzymes then rush in to promote the formation of a replacement cell. All this activity is what we call sunburn. The process of clipping out cells and growing new ones causes redness, pain, soreness, and peeling. People suffer this pain, make themselves look prematurely old, and risk cancer for the sake of looking good.

If the enzymes fail to detect the damaged cell, it and its descendants may hang around for years. When something finally activates the cells, they start dividing, and they produce the cancerous cells we all fear. What activates them and when they become activated remain medical unknowns.

Staying out of the sun is the best way to avoid skin cancer, but for most people, avoiding the sun completely is both unacceptable and impossible. Sunscreens help reduce risks, but they provide less than complete protection. The solution involves using common sense. Realize how the body protects itself, and avoid rushing things. The body protects itself by tanning, but tanning occurs several hours after exposure. It only makes sense to give the tan a chance to develop before getting a lot of sun. Start slow. Get a few minutes of sun, go back inside, and give the tan time to develop.

Authorities say most skin cancers develop from bad burns, not from the same amount of sun in several small burns. Children and teenagers are particularly susceptible to damage that later develops into skin cancer. Part of this could be due to the fact that teenagers devote more effort to suntanning than older people.

Skin cancers seldom develop in the places exposed to the most sun. Those areas have been exposed on prior occasions, and have a better tan than less-exposed regions like the hairline, the underarm, and around the eyes. Because they receive less sun, these areas are vulnerable.

Remember, the current decreases in ozone will cause few increases in skin cancer for 20 or 30 years. A few people have claimed that the current increase in skin cancer is evidence of ozone destruction.

That is unlikely. In my estimation, the current increase happened because people began many years ago to wear fewer clothes and spend more time in the sun—long before CFCs affected ozone.

As in everything else, there is a controversy between old and new experts when it comes to melanoma. The fact that neither side can prove the other side wrong says something about how poorly we understand ultraviolet damage. The older experts suggest that UV-B causes melanoma, and sunscreens provide protection. That is the currently accepted view.

An alternate view comes from two brothers, Cedric and Frank Garland. Cedric is an M.D. and the Chief of Epidemiology at the University of California-San Diego. Frank is a Ph.D. and head of the Epidemiology Department at the Naval Health Research Center in San Diego. These brothers claim that sunscreens indirectly cause increased cases of melanoma. The Garlands say that sunscreens fail to screen out the proper wavelengths.

The Garlands claim UV-A, which penetrates deeper into the skin, causes melanoma. Reference to Figure 1 shows that UV-A has longer wavelengths than UV-B. Sunscreens only partially protect us against UV-A. If UV-A causes, or even enhances, the cancer-causing ability of UV-B, then the use of sunscreens could contribute to increased melanoma, because people wear sunscreens and feel free to go into the sun. After all, they think they are protected.

Frank and Cedric Garland's primary argument involves sunscreen sales. They claim that melanoma increases are greatest in those countries where sunscreen sales are greatest. A counter to that argument could be made, however: that sunscreen sales are greatest in locations where people are most susceptible to melanoma. Most people recognize the risk and buy sunscreen, but not all people are so prudent. Some lie in the sun in the susceptible regions without sunscreen.

The sunscreen sales argument also claims that increases in melanoma since the 1960s have paralleled increases in sunscreen sales. This also seems a weak argument, because it is just what anyone would expect. If someone down the street develops a problem, most people on the street start taking precautions to make sure they avoid the problem. The result could be increased sales of sunscreen as melanomas pop up from unscreened exposures in the past.

For those who worry about the Arctic region developing ozone holes large enough to increase skin cancers in the United States, there is good news. The ozone hole would only last as long as the polar vortex. That means the hole might last from February through early March. Few people go to the beach or sunbathe during that time. Those involved in outside sports are probably skiing. Anyone who seriously worries about skin cancer would stay out of the mountains and avoid skiing, even when there is no ozone hole. Although skiing takes place below the stratospheric ozone layer, UV-B levels are much higher in the mountains.

THE ANTARCTIC FOOD CHAIN

Antarctica is a big, lonely place. Although its land area forms one of the largest continents, the only people who live there are nomadic scientists who migrate in and out annually. The only large animal is the penguin, and it never waddles far from the coast. There are no polar bears, seals, or walrus. At one time, it was popular to say not even germs lived there. What would they live on?

The ozone hole is large, just about as large as the continent. In some places, the hole spills over the coast, but it is generally the size of the ice-covered land region. The continental interior is devoid of life. Even migratory scientists seldom go there.

If an ecological disaster has to occur, Antarctica is the best place to put it. Ecological disaster may not even be a proper term for what could happen inside the Antarctic continent. Ecology deals with how living systems operate and interact. There are no living systems there except for the nomadic scientists, so speaking of the ecological system within Antarctica is sort of like talking about the ecology of the moon.

The coast, though, does have an ecological system. It is full of life, but the system is simple. The food chain starts with single-celled plants called phytoplankton (some plankton are animal and some are plant), ascends through small, shrimp-like animals called krill, and moves on to giant creatures like whales. That is the point at which most people get interested. We like whales. Plankton and krill evoke few feelings.

As things get closer to us in the evolutionary chain, we begin to care. Intelligence seems to be the key. The more intelligent a creature, the more we care what happens to it. We regard the fate of krill with indifference, but whales are intelligent, warm-blooded, and nurse their young like we do. In Antarctica, whales migrate in and out, rather like the scientists.

Except for whales, the things going on in the Antarctic ecological system have nothing to do with the rest of the world. None of the fish we eat come from there or depend on things that come from there. The Antarctic ecological system is almost a closed system. It is like a giant terrarium, shut off from the rest of the world. That is a major reason biologists like to study life there. It is a large but simple system, free from the interference of the rest of the world. It offers the opportunity to understand interactions between a limited number of creatures in an environment that hardly varies from year to year. Seasons change, but there are no droughts, floods, hot summers, or other complicating factors.

In an odd way, the ozone hole has been a boon to biologists who want to study life in Antarctica. Before the hole, there was little interest in the region. Now, the world waits to hear biologists talk about how the ultraviolet streaming through the ozone hole affects the plankton.

The Antarctic sea contains a variety of plankton. Some float near the surface; others stay a few meters below the surface. At any given depth, many varieties coexist. Early reports tried to assert that ultraviolet would be completely new to the plankton, like the smallpox Columbus brought to the Indians. That is not the case. The real worry is the sudden *influx* of ultraviolet. The hole may let as much ultraviolet in during the spring as normally comes in during the summer. In early spring, the only time the ozone hole exists, there was little ultraviolet before the hole. Thick layers of ozone typically give the polar regions more shielding than the rest of the world. In fact, the sun only reaches Antarctica in the spring by taking a long, sloping path through a lot of remaining ozone. Still, the incoming ultraviolet in the spring is an increase from no sunlight at all a few days earlier; and with it comes a lot more ultraviolet than was present before the hole occurred. After months of darkness, sunlight returns and brings a flood of ultraviolet instead of the usual trickle.

The effect on plankton has been likened to what would happen to untanned Norwegians who suddenly found themselves sunbathing in the Mediterranean during the middle of winter. Plankton have had no opportunity to build up defenses, to "tan" and protect themselves.

The idea of plankton "tanning" may sound strange, but they seem to do something like that. Deneb Karentz of the Laboratory of Radiobiology and Environmental Health at University of California, San Francisco, became one of those nomadic scientists going to the Antarctic. She wanted to find out how ultraviolet affects polar organisms. She reports that many make their own "sunscreen." Based on the 57 organisms she collected, Karentz concluded that 86 percent produce a built-in sunscreen, manufactured out of amino acids. Not surprisingly, those plankton living near the surface have better protection. Those living in deeper water need less protection, since the water itself filters out part of the ultraviolet.

"Phytoplankton have a wide range of responses to ultraviolet radiation," says Karentz. Some are exquisitely sensitive and can't survive any ultraviolet. She thinks some species may be wiped out and replaced by more resistant species as ultraviolet exposure in the Antarctic increases. We might call it survival of the fittest among plankton.

Ultraviolet can damage DNA or decrease photosynthetic efficiency. Plankton seem to repair broken DNA much like other living things. When bodies replicate the strands of DNA, even without ultraviolet damage, errors occur. Some sequences are wrong. In a process that sounds too miraculous to be true, enzymes move in, cut out the bad sequences, and replace them with new sequences. If the DNA in cells is damaged, as when people get sunburned, the enzymes still come to the rescue. The same process seems to work for plankton.

Experiments with land-based plants found that photosynthesis was retarded when light contained too much ultraviolet. No one knows how much is too much, but the obvious guess is that plants will behave much like plankton; some species will be unaffected, others strongly affected. Species that are unable to absorb sunlight and grow will decline in importance. Their decline opens a niche that other, hardier species will fill.

Much of the worry about ultraviolet wiping out the Antarctic life chain came from reports issued by Sayed El-Sayed, a phytoplankton ecologist at Texas A&M. He used filters to let in varying amounts of ultraviolet, while he measured photosynthesis by phytoplankton at Palmer Station in Antarctica. However, he thought he was increasing ultraviolet by six percent, when he was actually allowing a 50-percent increase. In response, photosynthesis decreased by one-half or three-quarters. El-Sayed now admits he made a mistake in his measurements, but he still predicts dire consequences. It is one thing to admit a computational mistake, as he does, but quite another to admit that your view of the future may be wrong.

At Scripps Institute of Oceanography, Osmund Holm-Hansen measured photosynthesis in Antarctica by putting incubation bottles at varying depths below the surface while the 1988 ozone hole was letting ultraviolet through. He reports that photosynthesis decreased by 15 to 20 percent in the top three feet of water, but the decrease "fell off exponentially" with increased depth. His conclusion: "The effect on overall primary productivity in the ocean in the southern hemisphere is fairly small. I don't see any scenario in which there would be a catastrophic effect. I am positive there will not be a collapse of the southern ecosystem."

Microscopic creatures adapt quickly. If some species stop functioning, others take over their nutrients and replace them in a matter of hours or days. Larger species do not adapt so quickly. If a disease wipes out the rabbits upon which coyotes rely for food, the coyotes are hurt. They starve, and their population declines. Several years pass before the coyote population comes to equilibrium. Microscopic creatures reproduce more rapidly and have more diverse populations, which allows them to adapt more quickly.

Nutrients limit life in the oceans. The Antarctic ocean teems with life because so much nutrient-rich water rises there. Incoming, nutrient-poor ocean water gets cooled in the Antarctic cold. That causes some waters to sink, and the displaced, nutrient-rich waters below to rise. Lack of nutrients is the only reason most of the world's oceans are not more productive.

Something, somewhere, will consume the nutrients that rise in the Antarctic. If the creatures now consuming those nutrients die off, others will grab the nutrients and thrive. If nothing was able to grow in the polar region because of increased ultraviolet, new life would spring up as soon as the flow of the nutrient-rich waters carried them out of the region affected by the ultraviolet. Life may be altered, but it will adapt. In the rest of the world, that adaptation might cause dislocations and hardship for people; on the other hand, it might create bonanzas of new food. Krill or other creatures in the Antarctic might suffer, but humans would hardly notice if the Antarctic ecosystem collapsed.

Sue Weiler, who once coordinated ultraviolet effects research at the National Science Foundation, summed the ozone hole up this way: "If the ozone hole had to develop somewhere, one would hope it would be over Antarctica." Her view is that the hole gives researchers the opportunity to examine the effects of ultraviolet in a setting where the impact is fairly small. "Antarctica provides a laboratory to help us predict what might happen if ozone depletion worsens in the Arctic and at temperate latitudes."

PLANTS OUTSIDE THE POLAR ZONES

Ultraviolet could damage or decrease photosynthesis in the Antarctic; it could do the same in nonpolar regions. The plants involved could be phytoplankton in oceans, crops in fields, or trees in forests. We may not be affected by the problems of plants in Antarctica, but the plants outside the polar zone are our food and our lives.

Plants protect themselves against ultraviolet. They need to block UV-B, at wavelengths between 280 and 320 nanometers (10,000 nanometers is about equal to the thickness of a sheet of paper), while letting the 400 to 1,000-nanometer wavelengths through. The latter band is the light needed for photosynthesis. When this band of light is diminished, photosynthesis decreases.

Plants subjected to increased ultraviolet respond by increasing in their tissue the levels of a kind of chemical called bioflavinoids. These flavinoids serve as sunscreens for the plant. Sometimes the flavinoids are interesting in and of themselves. For example, in marijuana plants, the protective chemicals are called cannabinoids, and these are the weed's main psychoactive ingredient. Growing marijuana indoors under ultraviolet lights might therefore increase its potency.

The effects of DNA damage are less important for large plants than for single-celled plants. Defective DNA in a single-celled plant means the whole plant is defective; larger plants can generally survive a few defective cells. Since DNA damage is of minor importance for larger plants, diminished growth efficiency is the thing to worry about.

A fairly large set of results is now in, demonstrating how plants—specifically soybeans—deal with increased ultraviolet.

In reviewing those results, each of us can see what we want to see. Those who anticipate bad news note that some soybean yields are down 25 percent. Those who prefer good news can focus on another variety of soybeans and point to no loss in yield whatsoever. So it goes with other plants, as well.

Alan Teramura, an ecologist at the University of Maryland, measures ultraviolet sensitivity in plants. He grows the plants in greenhouses with lights that increase the ultraviolet by 16 or 25 percent. What his results are seems to depend on where you read about his work. The difference between magazines is startling. Two are quoted here. One selection comes from *Discover,* September 1989. The other comes from *Reason,* June 1992. First is the quote from *Discover:*

"Soybeans, the third most important food crop in the United States, and loblolly pines, the source of two-thirds of our wood pulp for paper manufacturing, are among the plants that suffer. Too much ultraviolet light damages their DNA, destroys their chlorophyll, and disrupts photosynthesis, leading to stunted growth and a decrease in yield. Soybean harvests, says Teramura, would be down 20 to 25 percent if ozone levels dropped by 25 percent. Loblolly pines might take so long to reach market size that they would no longer be a viable source for timber."

Another view of Teramura's work emerges in *Reason:*

"Some crop varieties are sensitive to UV, so lower yields could result. For example, Teramura found a 25-percent reduction in yield after exposing one very sensitive variety of soybeans to a UV level corresponding to a 16 percent decrease in ozone."

Alarmists repeatedly cite Teramura's findings as evidence of the dire effects we can expect from a thinner ozone layer. But they fail to mention his concurrent discovery that several types of soybeans actually *boosted* their yields under increased UV, while others were unaffected. From *Reason:*

'Teramura has discovered large variation in UV sensitivity among different types (cultivars) of soybeans, corn, rice, and wheat. He tested 100 cultivars, including 40 types of soybeans, and found that 41 were unaffected by or tolerant of UV."

The truth may lie somewhere between these two media accounts. Soybean varieties are bred to deal with the environment in which they grow and to maximize yield. Perhaps the variety with the largest yield suffered a 25 percent loss from ultraviolet. Perhaps the varieties that showed increases in yield had negligible yields in the first place. The full story is longer and more complex than readers of either magazine would read. That is somewhat understandable. "Summarizing" the story is the norm in reporting. There is never enough room or reader patience, it seems, to do more. However, the reporter's obligation is to inform the reader, not to propagandize him. Summarizing the story is different than telling only the aspects of the story that lead the reader into agreement with the reporter's viewpoint.

Environmentalists, reporters, politicians, and scientists—all sides are guilty at times of bending the truth in the propaganda of the environmental war. Each side promotes its own version of the truth. When we begin to understand this, we're left with one haunting question:

What is the real truth?

2

Ozone and CFCs

Does chlorine monoxide have a right to eat, too?

In guessing how much global warming is really in store, the truth is this: Uncertainty abounds. However, that is not the case in tracking the cause of the loss of ozone. A few dissenting voices still ring out, but little doubt remains that CFCs are the major cause of the ozone hole. Some room remains to argue whether volcanoes or the space shuttle also contribute to the destruction, but not much.

The remaining arguments concern how large the hole will get, whether a hole will develop at the North Pole, and the vital question of how much ozone levels will decrease in nonpolar regions. Answers to the first two questions are apparently solid, but the vital one—how much ultraviolet levels will increase in nonpolar areas—remains open to debate.

Debate may not be too meaningful at this time. Strict cutbacks in CFC production are already underway. Our actions caused the hole, and we admit our guilt by cutting off production. All we can do now is live with what we have done. Immediate acceleration of the CFC phaseout would have little effect on future ozone decreases, because the damage is already done. The fact that we have released such massive amounts of CFCs in the past makes the remaining few dribbles insignificant. It is somewhat like the fellow who finds out he has lung cancer: Smoking one more cigarette is not going to make much difference.

Understanding why the last dribble makes little difference requires more explanation. Releasing CFCs into the lower atmosphere has no effect on ozone until the CFCs rise into the stratosphere. That takes over ten years. Even if we stopped all releases today, the concentration of ozone-destroying chemicals in the stratosphere would not immediately decrease or even stay the same. The levels would continue to increase for 10 to 20 years. It takes that long for the chemicals already in the lower atmosphere to leak into the stratosphere. The difference in future ozone levels between the currently scheduled phaseout and an immediate cutoff is negligible. The fellow with lung cancer smoking a cigarette may look silly, but that cigarette will have little effect on his future health. Fortunately for the world, the outlook is not so bleak as it is for the victim of lung cancer.

Everyone who argues about global warming should treat CFCs and ozone as an object lesson. Scientists argued about the connection between ozone destruction and CFCs for years before someone proved that the connection was valid. Once proven, it was too late to prevent the destruction from running its course.

In a sense, Mother Nature pulled a fast one. Destruction started in a place no one had ever suggested, and in a way no one had imagined. The setup required to make ozone destruction happen is so specific and unusual that, even with hindsight, it seems almost unbelievable. But it did and does happen. We must apply this experience to global warming and other environmental concerns. Once someone finds a substance or process that poses a potential environmental problem, we should admit that we cannot wait until we prove that the environment has been harmed.

Present phaseout rules for CFCs mandate that production must stop by the end of 1995. The only exception involves some production that can be maintained for essential uses and for servicing a few types of existing equipment. These special exemptions may mean that production continues at 15 percent of the 1986 production levels. Production levels in the U.S. are already less than half of those levels. Future levels of CFCs in the atmosphere, however, depend on releases of CFCs manufactured years ago, and on CFC releases that occur when equipment breaks, is salvaged, or plastic foam is recycled.

Consumers have some strange ideas about who pays the cost of phasing out CFCs. Many think that the phaseout hurt chemical companies, but the main concern to chemical companies is whether they can hold onto their old customers when new CFC replacements reach the market. Companies worry a lot about keeping customers and about profitable product lines that disappear.

However, consumers will be the ones paying the direct transition costs, not the companies. Consumers have a giant investment in equipment that depends on CFCs. The substitutes may work fine, but not in the consumers' old equipment. Replacements for CFCs may demand special operating conditions—things like higher pressure lines, special lubricants, and larger heat exchangers—in other words, newly designed equipment. Replacing the refrigerator or deep freeze rather than repairing it might be necessary. The same applies to household and car air conditioners. And the replacement equipment will be more complex, which means it will probably cost more. CFC replacements are unlikely to be as energy-efficient. That means higher utility bills. To effectively cut down on future CFC releases, regulations would have to require reclamation of CFCs from old equipment. You may have to pay $100 to get rid of your old fridge.

According to estimates, people in the U.S own $135 billion worth of equipment that depends on CFCs. Most people imagine that the equipment consists mostly of machines belonging to industries, and the average consumer does not care too much if industries have to pay the cost of replacing equipment. But consumers are figuring wrong—they themselves own most of this equipment.

Calculate this. Many homes have a refrigerator, two automobile air conditioners and a home air conditioner. These have a combined replacement cost of at least $2,000. If 50 million homes have to replace these items, the replacement cost amounts to $100 billion. Obviously, the consumer has a big stake in this process.

The longest portion of the delay in introducing CFC replacements was not primarily due to the companies' reluctance to start making new products. It was due to the time required to test the toxicity of the CFC replacements. Testing takes five years and costs so much that chemical companies formed consortiums to handle the testing. This helped cut costs and may have speeded up the testing.

OZONE LEVELS OVER TIME

All the debate about ozone levels and their importance makes it seem that we know a lot about their history. But that history is amazingly short. Before the 1970s, ozone records can only be described as scattered and spotty. Before the 1950s, records are virtually nonexistent. If ozone levels had been constant and unchanging through those times, the decrease now would be startling. However, they have not been constant. Ozone levels change from day to day, and show sizable swings throughout the existing record.

Those doubting the ozone-CFC connection question whether the chlorine oxide from CFCs causes the ozone hole. Their argument is that perhaps ozone holes simply occur from time to time, and now happens to be one of those times. With so little known about the variation of ozone, that might be a valid argument. In fact, evidence suggests that low ozone levels did occur in Antarctica during the 1950s.

However, proof of the ozone-CFC connection comes from measurements inside polar clouds, not from ozone history. Connections between CFCs and ozone decreases outside of the polar region remain controversial. Discussing ozone variations next will give us some idea of the degree of changes the world regularly survives, and will provide some perspective on how large the decreases must be before they are harmful or even meaningful.

Figure 1 offers an example of ozone variation. It shows the total ozone above New Delhi, India, from the 1950s until 1980. During most of this period, say until 1970, any effects from CFCs were small. The ozone is measured in Dobson units because G. M. B. Dobson came up with a spectrometer that could determine the total amount of ozone above the observing site. Measurements could be made anytime during daylight hours, even when clouds

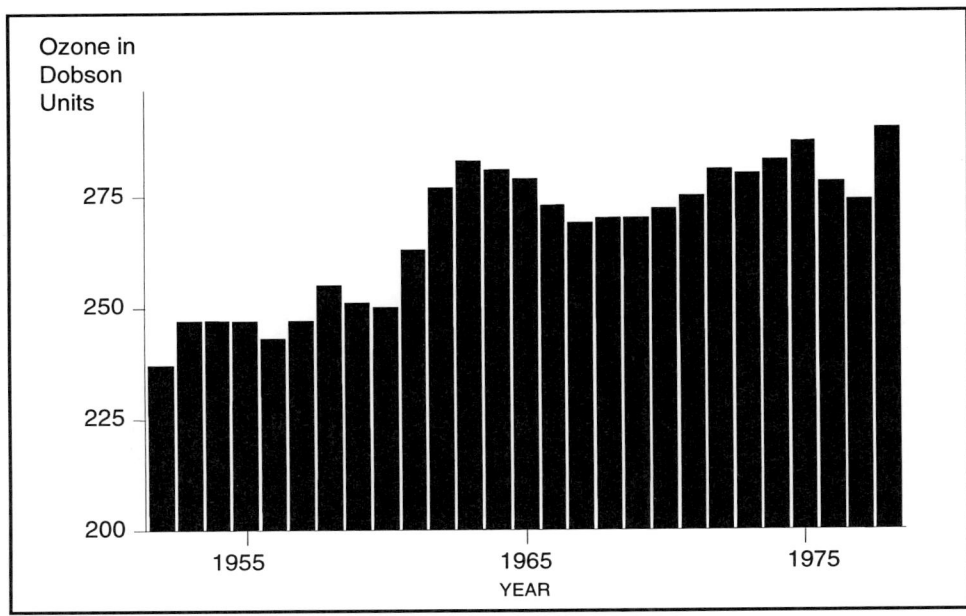

FIGURE 1 *The figure shows the amount of ozone measured over New Delhi, India. Overall, the ozone showed a ten-percent increase during this 30-year period. The figure also illustrates the variability of ozone levels.*

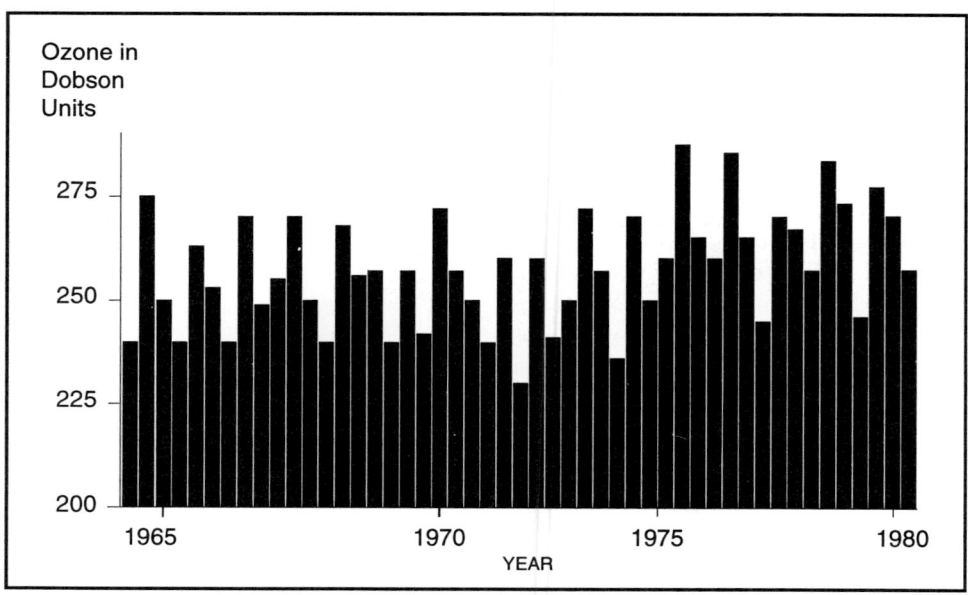

FIGURE 2 *These ozone values were measured at Ahmedabad on the Indian subcontinent, but are presented with a shorter averaging interval than is used in Figure 1. The shorter interval reveals the annual variability. Each year, ozone increases and decreases by about ten percent. The same pattern occurs in most of the world.*

covered the sky. It provided an inexpensive and reliable way to track ozone variations. Of course, no one had any good reason for tracking ozone variations until recently. Few scientists devoted themselves to filling books with data on ozone, because it was data that had no immediate or known usefulness.

In 1992, newspaper headlines warned us of a three-percent loss in ozone over the last decade. In the decade before 1965, New Delhi saw almost a 20-percent increase—larger than most places experienced, but not by much. The day to day variations in ozone are much larger. Let's say the jet stream shifts as a storm moves through. The shift might cause a 50-percent rise or fall in the ozone above sections of the United States.

Ozone undergoes yearly oscillations and shifts as weather patterns change. A dozen points had to be averaged together

to arrive at the New Delhi data. Figure 2 illustrates the extent to which ozone varies on a month to month basis. This data came from Ahmedabad, another measurement station in the Indian subcontinent.

Curves such as these make you wonder how a decrease as small as three percent could even be measured. In fact, the words "three percent loss per decade" imply a smooth, uniform process. In all likelihood, the current average data from several stations gave an answer that was three percent less than the same average ten years ago. With no long-term data to show what the true average ozone level is, not much else can be done. Without the data, who can say that three-percent variations over a decade are not expected? In fact, most experts say they *are* expected. In the following decade, ozone levels may be back up.

Scientists are trying to detect ozone

depletion without knowing what causes natural variations. It is as unfair to contend that all decreases are due to CFCs as it is to wait until a natural increase occurs, then say that the increase proves that CFCs do not destroy ozone (strictly speaking, CFCs do not destroy ozone, but the chlorine produced when CFCs break up can do so).

In the spring of 1993, newspapers neglected to mention a decrease much larger than three percent per decade. That decrease was nine percent, and it occurred in about a year's time. CFCs had nothing to do with it. No one quite understands how, but the loss seems due to the volcanic eruption of Mt. Penatubo. Such a large loss during a single year illustrates the difficulty of establishing long-term trends.

Just what causes natural variation in ozone levels remains a mystery. It appears that ozone varies with what is called the solar cycle. The causes and effects of this variation are not well understood. Whatever other factors cause ozone to vary are even less understood. Our knowledge of what causes ozone to vary is, in fact, so poor that no one tries to predict future variations—except for the predictions of decreases due to CFC-related ozone destruction.

What do we know about previous ozone holes? Nothing. Some evidence suggests that South Pole ozone levels were quite low in the 1950s. This evidence comes from three sources—two sets of direct measurements and an indirect method for deducing past ozone levels. Two French scientists recently republished data showing large ozone decreases in the Antarctic during the spring of 1958. Ozone levels of over 300 Dobson units would be normal, but these researchers found levels of only 120 Dobson units. Since the scientists have only one set of measurements, someone might say, "Those guys did

not know how to use the spectrometer properly." But the other set of measurements overcomes that objection. It was taken by G. M. B. Dobson, the inventor of the technique and the man that the ozone units are named after. He measured levels of only 150 Dobson units over Halley Bay, Antarctica, in 1956 and 1957.

The indirect set of measurements needs a little more explanation. Plants protect themselves from ultraviolet light by producing bioflavinoids. Kenneth R. Markham of New Zealand's Department of Scientific and Industrial Research took advantage of this phenomenon when he was analyzing collections of herbarium moss. Researchers collected the moss in the Ross Sea area of Antarctica between 1957 and 1989. To protect itself from ultraviolet radiation, the moss produced more flavinoids during years of reduced ozone levels. Measurements of flavinoid levels in moss during the 1980s, after the ozone hole appeared, verify the relationship between flavinoids and ozone. As the ozone hole developed, the moss produced more and more flavinoids. The moss also produced high levels of flavinoids in 1958. This suggests that 1958 was a year of high ultraviolet, and therefore low ozone. The next available moss sample was in 1964, and ozone levels had apparently recovered by then. Flavinoid reflection of the ozone levels provides only qualitative evidence, so the exact number of Dobson units of ozone cannot be calculated.

These results raise the possibility that the ozone hole could be caused by things other than CFCs. If an ozone hole occurred in 1957, it would have little to do with the CFC problem. Such a hole back then would indicate that something else might also cause the problem. It would provide a reason to look for the something else, but not a reason to forget about CFCs. The existence of an ozone hole in the late

1950s would in a way be reassuring. It would mean the hole came and went without creating problems serious enough to notice. It is possible that the current hole will also go away and leave no major problems behind.

OZONE SCARES OF THE PAST

Worries that the world might lose its protective layer of ozone started over 20 years ago. Then, CFCs were not accused as the culprit. Reviewing all the things that someone claimed might destroy ozone and understanding why they did not do so will demonstrate how erratically science moves. Lack of ideas, not lack of money, limits the rate of progress. The on-again, off-again attempts to regulate CFCs show how difficult it is for regulators to decide what poses a hazard and what to do about it. This history also reveals the problems of industry. As government sent signals that regulations were coming, producers started large research programs to find replacement products. As the government canceled plans for regulations on CFCs, companies canceled research programs for replacements.

Essentially, the CFC-makers acted reasonably throughout this difficult period. In hindsight, it is too easy to complain that the manufacturers should have stopped production on the basis of the first scares. Notice that the early scares were all false alarms. At that time, industry was right when it claimed that the case against CFCs was unproven. In the end, CFCs do lead to ozone destruction, but not in the ways first suggested. The early case against ozone remained unproven because it was wrong.

In 1970, a national debate raged. The British and French were developing a supersonic jetliner, and U.S. companies wanted Congress to provide funding so we could develop a competing supersonic transport (SST). Proponents argued that

SSTs were the aircraft of the future. They claimed our aircraft industry would be doomed if we did not develop our own SST. Opponents claimed that SSTs cost too much, that they were not in fact the wave of the future, and that companies should pay to develop them, not the government.

By raising environmental concerns, opponents came up with the trump card that defeated SST development. SSTs were designed to fly high in the stratosphere, and they would release nitric oxide as they flew. Someone suggested that this nitric oxide would destroy the ozone layer. This was the first time most people had ever heard of the ozone layer. Howard Johnson at the University of California-Berkeley calculated that 500 SSTs, flying seven hours a day for ten years, would wipe out 22 percent of the world's ozone. No one, not even Congress members themselves, can say why Congress does anything, but after much talk about ozone levels, Congress killed SST funding in 1971.

Stung because they had been unprepared for the ozone destruction question, staff members at the Department of Transportation established a Climatic Assessment Program to warn us of any more surprises concerning planned aircraft and space launches. NASA had the space shuttle under development, and the panel next pointed an accusing finger at the shuttle. Its reusable rockets would release hydrochloric acid and other chlorines as the rocket burns. The panel said chlorine could destroy ozone as easily as the nitrogen oxides from the SST. However, it seems that everyone favored our space program, so no one paid any attention to the panel.

Given hindsight, we know that hydrochloric acid had no ability to destroy ozone through the reactions suggested at the time. For several years, in fact, science claimed that chlorine forming into hydrochloric acid in the stratosphere kept chlorine in CFCs from getting to the ozone. It does. The ozone hole develops when something interrupts the hydrochloric acid.

Since nitrogen oxides could destroy ozone and fertilizers are nitrates, it was only a matter of time before someone accused fertilizers of threatening the ozone. In 1975, Michael McElroy from Harvard concluded that increased use of chemical fertilizers had to stop. According to his calculations, ten percent of the ozone would be gone by the year 2015 if chemical fertilizer use continued to increase. The talk about fertilizer caused a commotion in scientific circles, but the public ignored the controversy. Fertilizer use continued, and that turned out to be the right public policy as far as ozone is concerned.

The fertilizer debate never got much attention because the public was more worried about something else that was destroying ozone. That something else was CFCs. Interest in CFCs developed in a roundabout way.

A Britisher by the name of James Lovelock had developed an electron capture detector that could detect extremely low concentrations of CFCs, and could easily measure CFC concentrations in ordinary air. Before the device was used, no one had noticed that CFC levels had become so high. Lovelock had no particular interest in CFCs, but he obtained funding to go around the world mapping CFC concentrations. He even got his funding without claiming that CFCs were a threat. Lovelock once described them as offering "no conceivable hazard."

People like Lovelock make funding science a nightmare. All too often, the path for important new findings is established by someone who is doing something without apparent justification, as Lovelock was doing.

Sherwood Roland at the University of California-Irvine heard of Lovelock's work and became curious. If so many CFCs were floating in the air, what finally became of them? Apparently no one had asked the question before. They were not toxic, did not cause harm by interacting with anything, so why ask? Release them, they fly away, and you never hear from them again. It seemed an ideal chemical.

Roland put a student, M. J. Molina, to work on the problem. The pair found that nothing in the lower atmosphere interacted with or absorbed CFCs. These gases could float around the troposphere forever, and nothing would interact with them. The only way they would ever leave the troposphere was by diffusing up into the stratosphere.

Once CFCs entered the stratosphere, the strong ultraviolet light could tear them apart and destroy them. This process made Roland and Molina recognize CFCs as a threat to the ozone. When the ultraviolet broke them down, they released chlorine, which could quickly destroy ozone. Chlorine's ability to destroy ozone had already been suggested by the Climatic Assessment Panel with reference to the space shuttle.

Now it is necessary to pause for a little background material. Ozone is not a finite resource like gold. With gold, there is only so much available and no more. With ozone, lots more is available. Solar ultraviolet creates ozone in the stratosphere all the time. If we destroy it, more is built. As with anything, the more ozone there is, the more rapidly it gets used up and destroyed. Nature keeps building ozone until the supply gets so large that destruction rates match creation rates. Increase the destruction rate and the supply goes down, but not to zero. The faster the destruction, the lower the ozone levels become.

Ozone is not complex. It consists of just

three oxygen atoms joined into a single molecule. Oxygen molecules in the air are made up of two oxygen atoms joined together. Add one more, and you have ozone. Here is how the additional atom can be added. Solar ultraviolet splits an oxygen molecule into two atoms. Either of these newly freed oxygen atoms can hook up to an oxygen molecule (two atoms) and transform it into an ozone molecule. Figure 3 illustrates the process. Figure 4 illustrates how nitric oxide, or NO, can destroy ozone. Notice that two reactions take place. The first one results in the

formation of nitrogen dioxide, or NO_2, and the second transforms nitrogen dioxide back into nitric oxide. This means nitric oxide can destroy ozone without decreasing the number of nitric oxide molecules.

This nitric oxide reaction happens constantly. It is the way that nature has always limited supplies of ozone in the stratosphere. The problem arises when people increase the supply of nitric oxide to levels much higher than nature provided. That was what users of chemical fertilizers were allegedly doing.

With some understanding of the way

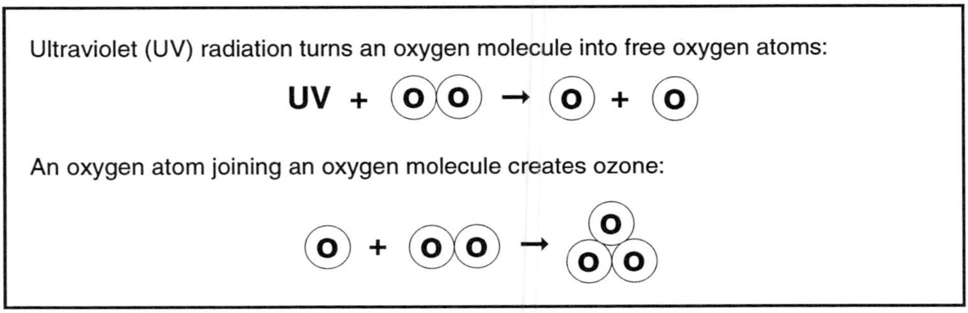

FIGURE 3 *The upper section shows how the ultraviolet in sunlight tears oxygen molecules apart to form free oxygen atoms. The lower section shows the free oxygen interacting with oxygen to form an ozone molecule. Ozone is nothing but three oxygen atoms joined to form a molecule.*

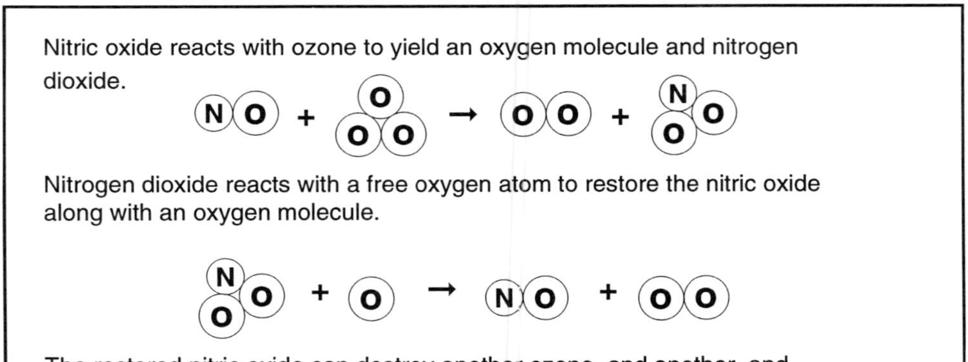

FIGURE 4 *The first concern to emerge in connection with ozone destruction involved nitric oxide generated by high-flying aircraft. In the first reaction, nitric oxide destroys a molecule of ozone and becomes nitrogen dioxide. In the second reaction, the nitrogen dioxide gets converted back into nitric oxide, ready to destroy another molecule of ozone.*

UV from the sun causes CFCs to release chlorine.
Chlorine destroys ozone to make chlorine oxide and a molecule of oxygen.

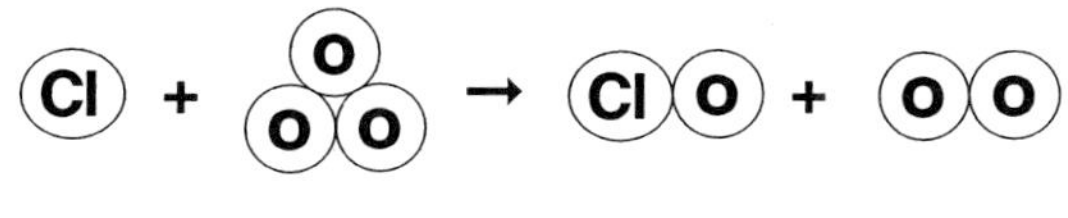

Chlorine oxide and a free oxygen atom react to make chlorine and a molecule of oxygen.

Since the last reaction freed a chlorine, that chlorine can keep repeating this same reaction, destroying one more ozone molecule in each cycle.

FIGURE 5 *A single chlorine atom can destroy unlimited numbers of ozone molecules, because the chlorine gets recycled. Chlorine becomes chlorine oxide when it destroys a molecule of ozone. A free oxygen atom then converts the chlorine oxide back into chlorine, and it can now destroy more ozone.*

nature makes and destroys ozone, Roland and Molina's ozone destruction theory is easy to explain. When solar ultraviolet breaks CFCs apart, one part released is chlorine. CFC stands for chlorofluorocarbons, with the "chloro" part representing chlorine. The CFCs themselves offer no threat to ozone; CFCs do not react with many things. It is the chlorine portion of the CFCs that eats ozone. Figure 5 shows how the meal goes, and goes, and goes. This is another of those reactions where the chlorine gets restored so it can eat just one more molecule of ozone, restored again so it eats just one more, and so forth—forever.

Chlorine is not the only ozone killer released in large quantities. Halons are even more effective destroyers of ozone. Halons might be considered cousins of CFCs, since the only difference is that halons contain bromine instead of chlorine. Estimates report that halons account for 25 percent of the ozone loss; CFCs for the other 75 percent. Measurement programs have concentrated on chlorine and neglected bromine, but experts believe that bromine and chlorine behave in the same manner. The discussion that follows concerns CFCs, but you can take CFCs to mean both CFCs and halons, since they act in much the same way.

For some reason, every article on ozone destruction employs the same cliché: "Each chlorine atom can destroy 100,000 molecules of ozone." The refrain implies that the chlorine atom eventually gets tired and wears out. However, atoms do not work that way. They keep on reacting forever. What the cliche refers to is that the chlorine eventually gets carried out of the stratosphere or becomes permanently locked up in a nonreactive molecule. Only in this manner will it stop destroying ozone.

Roland and Molina published a paper in December of 1973, predicting that CFCs would lead to a drop in ozone levels of between seven and 13 percent by the end of the next century. At the time, world production of CFCs was almost two billion pounds per year—about a half pound for each person in the world. Production was doubling every five to seven years. Aerosol spray cans contained half of this production, and we used more than our share in the U.S. The prediction that CFCs would destroy so much ozone really shook the country. The public reacted by cutting way back on its use of aerosol cans.

People also wanted Washington to stop production of aerosol cans, or "ban the can." The Senate could never summon up enough courage to take such a definitive action. One bill to ban the can received only 26 votes. Instead, Congress passed the buck to the Environmental Protection Agency. The EPA was directed to ban aerosol cans if there was evidence—not necessarily proof—that the cans were harmful to the environment. The EPA did ban the can in 1978, but use in the United States had already halved. Most of the world ignored the potential problem. The only other countries to ban aerosol cans were Canada, Norway, and Sweden.

The National Academy of Science was directed to report on ozone destruction. The Academy had circulated a draft report saying CFCs would cause losses of up to 14 percent of the ozone—about the same percentage Roland and Molina had estimated. Then, in February of 1976, Roland and Molina published another result. By this time, the pair realized that most chlorine would be put out of action by other molecules, called reservoir species. One such species was hydrochloric acid; another was chlorine nitrate. This caused the researchers to significantly reduce their estimate of ozone losses. The National Academy of Science also revised its report before releasing it. A new estimate projected a seven percent loss. Actually, the estimate indicated losses between two and 14 percent, which was quite a range.

This new result did not clear CFCs; it only halved the ozone loss expected. Still, the public had been worrying about ozone

for three long years. It was time to move on to some other crisis. Everyone, including the EPA, seemed to lose interest in CFCs.

Another result materialized during the summer of 1977. A study found that some reaction rates involving nitrogen oxides occurred at a rate 40 times faster than what was previously supposed. This result meant that SSTs would not have harmed the ozone, since the nitric oxide would have quickly reacted with something else. More importantly, it meant that the reservoir species could corral less of the chlorine than expected. The National Academy of Science took this result into account and more than doubled its previous estimate of predicted ozone losses—up to 16 percent.

The Academy strongly advocated a coordinated, international strategy for the control of CFCs. Disagreement came across the sea from the British Stratospheric Advisory Committee. The committee reported that its computer models predicted the same amount of loss, but the scientists there did not trust their models. Industry took heart from the British view and stated that the case for ozone depletion remained unproven.

The public was understandably confused by this fluctuation in the estimates. Besides, energy worries had now superceded ozone worries. When, in 1980, the EPA published proposed restrictions on CFC uses in the Federal Register and invited comments, over 2,000 comments were received. Only four comments supported the control measures. The EPA abandoned the restrictions.

Industry subsequently stopped research on replacements for CFCs. In fact, companies developed new uses for CFCs. This was consistent with actions taken by the National Academy of Science. In 1982, the Academy released another report. This one again reduced the estimate of ozone losses.

The new estimate put ozone losses at between five and nine percent. The world yawned, just as the reader may be doing by now. The Academy released yet another report in 1984, suggesting losses of only two to four percent. Few people paid any attention.

This, then, is the challenge of trying to regulate on the basis of new scientific findings. It sounds like a mess; it was, and it is. There are no bad guys here, just confused players. Trying to pass regulations too soon after new scientific findings is a mistake. Findings have to "cure" for a few years before other scientists have much faith in them. After the ozone fiasco, the public would be justified in wanting to wait for definitive results. Both science and the public might have been well-served if all this had stayed out of newspaper headlines and off nightly newscasts at the time.

OZONE DESTRUCTION IN ANTARCTICA

Early scares about ozone depletion throughout the world fell flat because the chlorine kept getting itself bound up in reactions that prevented it from affecting ozone. Chlorine was supposed to destroy molecule after molecule of ozone without restriction. When other reactions intervened, the chlorine and could no longer destroy ozone.

Two reactions saved us from early, worldwide ozone destruction. Methane molecules react with chlorine to form hydrochloric acid. Chlorine oxide also gets captured in a reaction with nitrogen dioxide that produces chlorine nitrate. Each of these reactions captures a chlorine atom and keeps it from doing any mischief to our ozone. The chlorine is still up there, but it is locked away in what scientists call reservoir species.

It was due to good fortune, not good management on the part of humans, that such reactions took place. Otherwise, much of our nonpolar ozone would already be destroyed. Once we discovered these reactions we felt protected, so more uses were found for CFCs, and production expanded.

Most people were complacent until 1985, when Joseph Farman of the British Antarctic Survey announced the existence of the ozone hole to the world. Farman published measurements taken at Halley Bay in Antarctica, showing that the ozone had decreased from above 300 Dobson units in 1968 to 200 Dobson units in 1984. One wonders why Farman did not publish his findings in, say, 1980, when the ozone had already dropped to 230 Dobson units. Obviously, he thought such drops were just part of ozone's normal variability.

Stories of scientific discovery often sound like cases for Sherlock Holmes. A mystery exists until some scientist steps forward with an obscure piece of information that resolves the mystery. The piece of information is so specialized and of such small interest that no one else has bothered to absorb it. That is how the explanation of the ozone hole came about. It had to do with clouds in the polar stratosphere.

The only place and time stratospheric clouds can exist is in the extreme cold of the polar winter. The stratosphere is too dry to form stratospheric clouds anywhere else. Temperatures have to drop to -181°F before water condenses out of such dry air. At such cold temperatures, ice crystals form instead of water droplets. Even in the cold of the polar regions, such clouds are rare, occurring for only a few days during each polar winter. This is where the Sherlock Holmes piece of the puzzle comes in.

In 1983, Patrick McCormick at NASA had discovered evidence of polar clouds in satellite data. However, the clouds occurred where temperatures were too warm for water vapor to condense—as warm as -172°F. Science is full of odd observations like this one. Contrary to events in science fiction movies, the world of science does not come to a halt in order to explore the new anomaly. Most scientists never hear of the results, and those who do assume the reports are false alarms due to bad measurements. Every so often, though, someone recalls an old anomaly, just like Sherlock Holmes, and explains something important. That is what happened with the polar clouds. Clouds and ozone destruction have no natural connection. A long chain of events is needed to forge the link.

The first nebulous clouds appear on sulfate aerosols. Sulfates start out as sulfur dioxide gas from volcanoes, coal-burning power plants, and several other sources. Something—and no one is sure what—causes the gas to condense, and it turns into a cloud of aerosols. As temperatures cool, water can condense onto these sulfate particles. Water from the condensation dissolves the sulfates to form microscopic droplets of sulfuric acid, much smaller than most cloud particles. These clouds do not affect the ozone or the chlorine reservoir species. Instead, they are prerequisites to the formation of another kind of cloud.

Gases cannot condense without some form of nuclei to settle on. Serving as nuclei is the role of the sulfuric acid clouds. Nitric acid condenses on the sulfuric acid droplets, and these grow to produce a stratospheric nitric acid cloud. The processes do not occur everywhere in the polar region simultaneously, nor do they form large clouds. Such clouds form in clumps and bunches, just like water clouds. Their location is determined by air circulations far below, in the troposphere. These are the clouds Patrick McCormick had seen in satellite data. Nitric acid

clouds can form at the "warm" temperature of -172°F, a temperature too warm for water (ice) clouds to form.

Ozone destruction occurs like the nitric acid clouds, in clumps and bunches. Ozone destruction within the hole varies from place to place, just as the clouds vary.

When Mt. Penatubo erupted, there was fear that the aerosols and sulfates it injected into the stratosphere might work like polar clouds and destroy a lot of ozone. After all, it did provide sulfuric acid droplets and aerosol particles. No immediate decrease occurred, though, so scientists found reasons to rationalize why such losses did not occur. Then, a year later, satellites revealed significant ozone losses over the U.S. and mid-latitudes after all, amounting to about nine percent. Compared to the three percent per decade loss expected to result from CFCs, this is a giant loss.

It's doubtful that the sulfates and aerosols directly caused the loss. If they had, we would have seen the loss right away, instead of spotting it after a year's build-up. Penatubo caused surface temperatures to decrease and stratospheric temperatures to increase. The aerosols it injected into the stratosphere absorbed sunlight, causing the stratosphere to warm. This absorption decreased the sunlight reaching the ground. Less sunlight at the ground meant cooler temperatures.

The best guess about what caused the ozone to decrease has to do with changes in air circulation within the stratosphere. Heat itself has little effect on ozone. Also, there was no ozone loss over the equatorial regions, only at mid-latitudes. Here is a case of a very large ozone loss taking place in a way that scientists never visualized. The loss caught them and their models unaware.

Water clouds sound peaceful, fluffy, and safe, while clouds of harsh chemicals like sulfuric and nitric acid sound uncomfortable, dangerous, and alien. However, ozone itself is vicious stuff. Lungs become raw after only a few whiffs. Ozone may do desirable things in the stratosphere, but it is a primary irritant that causes burning lungs in polluted air.

Nitric acid clouds disrupt the reservoir species that normally keep chlorine from destroying ozone. Removing nitric acid gas from stratospheric air has the ripple effect of reducing other nitrates. One of those nitrates is the nitrogen dioxide that deactivates chlorine by forming chlorine nitrate. The loss of nitrogen dioxide should be no big deal, because the chlorine is already bound up in chlorine nitrate. If the chlorine is bound up, the loss of nitrogen dioxide seems unimportant. But it turns out that a conspiracy takes place, and chlorine is released. One other important point: Nitric acid cloud particles are small. They fall slowly—so slowly that they never get out of the stratosphere.

Sometimes a nitric acid cloud finds itself bumped down to altitudes where the weather gets really cold, say 181°F (stratospheric temperatures get colder as you go down instead of colder as you go up). That is cold enough for water to condense, and the nitric acid cloud particles serve as the nuclei that water needs before it can condense.

These are not the nacreous ice clouds seen in polar regions. Those clouds condense on some other kind of nuclei, nuclei that are smaller. Ice particles in nacreous ice clouds start small and never get too large. The nitric acid droplets serving as nuclei for water are almost as large as nacreous cloud particles ever get.

When ice particles start out this large, big ice particles can grow. At only tens of micrometers long, these are not big by snowflake standards, but they are large enough to fall a few thousand feet a day— fast enough that the ice particles can fall

entirely out of the stratosphere. Recall that the stratospheric supply of nitrates condensed into the nitric acid cloud particles, then ice grew on these particles. When the particles fall out of the stratosphere, they leave behind a stratosphere depleted of nitrogen dioxide, the molecule able to capture and hold the chlorine that destroys ozone.

As in a well-constructed mystery, other events took place as the ice particles grew. Water ice provides a surface where the two reservoir species, hydrochloric acid and chlorine nitrate, can get together. The two "mate" to produce nitric acid and release chlorine. The reaction can take two forms, but the net result is that nitric acid stays with the cloud particle and releases chlorine gas.

The chlorine gas is now free. All the nitrogen dioxide that normally sops it up has been depleted from the air by the formation of nitric acid cloud particles. There is nothing to keep the chlorine from destroying the ozone except the darkness. All this takes place during Antarctic winter, a time when nights are six months long and the sun never shines. Come springtime, the sun comes back. The chlorine sits waiting. Chlorine begins to destroy the ozone at altitudes between 30,000 and 60,000 feet. The altitude range may be getting larger as chlorine builds up and the hole gets deeper.

One other critical event takes place in the Antarctic each winter—the formation of the Antarctic vortex. The vortex is a sheath of air circulating around and around the pole, forming a barrier that lets no air in or out at the altitudes of the clouds. It is a kind of jet stream with winds as fast as 200 miles per hour. The vortex forms because winter air cools, creating a low pressure region. Air starts to flow into the low pressure region and gets deflected into a circular flow by Coriolis forces, the same forces that cause oceans currents to form gyres and hurricanes to rotate in circles.

In the Antarctic, conditions allow the vortex to last all winter. It does not break up until after the sun comes back in the spring. As a result, polar stratospheric air is isolated, preventing the chlorine from drifting away and weather fronts from bringing in new nitrogen dioxide. Without the vortex, the hole would never develop, because the air would remain mixed. Polar air would not cool off enough to form clouds, because warm air would enter and keep temperatures too high.

The polar vortex is larger than the Antarctic continent. At this point, the ozone hole has not filled the entire vortex. Why not? We're not sure, but there is little reason to doubt that the hole will expand until it fills the vortex. That means it may double in size over the next few years. Still, the vortex will confine it to the Antarctic region.

The ozone hole could in turn affect the time at which the vortex breaks up. Since ozone absorbs ultraviolet to warm the upper atmosphere, lack of ozone means the upper atmosphere warms more slowly. The vortex seems to be lasting a little longer each year. Normally, ozone absorbs sunlight, causing the air in the vortex to warm, and speeds the vortex breakup.

This is a beautifully complex chain of events. The complexity provides assurance that the hole cannot spread to cover the world or even to destroy all the ozone above the South Pole. Below the tropopause, the lower boundary of the stratosphere, temperatures, water concentrations, and circulations are different. These conditions form a floor that keeps the hole from descending much further. Higher in the stratosphere, temperatures become too warm for ice clouds to form. This puts a lid on the top of the ozone destruction region. There will always be some polar ozone, but maybe only one-fifth the normal amount. That amount is enough to provide some ultraviolet protection.

OZONE DESTRUCTION OUTSIDE OF ANTARCTICA

A polar vortex and stratospheric clouds also form at the other end of earth—the North Pole. This area, too, may include an ozone hole someday, but probably not on a regular basis or at the depth of the South Pole's ozone hole. Here is why:

The atmosphere is warmer over the North Pole than over the South Pole. As a result, fewer stratospheric clouds form there. However, clouds do form and measurements have revealed conditions that seem ripe for the creation of an ozone hole—not a big one, but a hole nevertheless.

Still, no hole has occurred. That is because the vortex around the North Pole is not as stable—not as airtight—as the one at the South Pole. Storms forming in Siberia disrupt the vortex, pull streams of air away from it, or send puffs of air into it.

Those disruptions are not too important during the midwinter. But as spring approaches, they cause the Arctic vortex to break up before spring arrives. It is the coming of spring that brings the sun and starts ozone destruction in the Antarctic. Without the sun, the chlorine just sits there and can destroy no ozone. As soon as the chlorine eats one ozone molecule, it finds itself bound up as chlorine oxide. It takes ultraviolet from the sun to break up the chlorine oxide and free the chlorine atom so it can swallow another ozone molecule. As long as the Arctic vortex breaks up before spring, there is no ultraviolet, and there can be no Arctic ozone hole. This statement needs some qualification.

Someday, the breakup may come late. If it comes too late, an Arctic ozone hole will form. But unless climate changes, the late breakup should be the exception, so the likelihood of a regular Arctic ozone hole is small.

What about loss of ozone in nonpolar regions? Without stratospheric clouds, there is no need to worry about ozone holes forming anywhere but at the poles. And as long as stratospheric temperatures remain at current levels, there is no chance that stratospheric clouds will form in nonpolar regions. The answer, however, has a built-in hedge. What if stratospheric temperatures change?

Despite the reassurance, we can't say that no ozone loss whatsoever will occur away from the poles. For starters, Mother Nature may yet devise another means of destroying ozone. No other way has been suggested, but the current problem was not suggested until Joseph Farman published measurements showing that a hole already existed. There is no way to know if some other human activity, something not having to do with CFCs, will be found to destroy ozone. Every time we do something new, we might possibly be setting off cataclysmic events. We must be careful, but there is no way we can know what is careful enough.

Events at the pole can also cause loss of ozone elsewhere. When air from the ozone hole mixes with other air, there is a net decrease in concentration. Even if no more ozone gets destroyed, some of the rest of the world's ozone moves into the hole, and air without ozone moves out. That process creates a net decrease in the amount of ozone that exists in the world. The loss may be noticeable in New Zealand or Southern Australia, since those areas are close to the pole. In other parts of the world, the increases should be so slight that we won't even be able to attribute them to the ozone hole. Currently, the Antarctic ozone hole accounts for three percent of the world's ozone. If the hole doubles in size, it could account for up to six percent. But not for the full year. The resultant decrease would only last days or weeks, until sunlight and oxygen created new ozone.

CFCs ARE GOING, GOING, ALMOST GONE

The ozone hole got the world's attention. Farman published his paper announcing the hole's existence in 1985. By 1987, a group of 27 nations had signed the Montreal Protocol, which limited use and production of CFCs and halons; the agreement represents fast work in diplomatic channels.

In hindsight, the Montreal Protocol does not seem as strong as it might have been, but it was a first step, and it effectively ended the argument about whether something should be done. It acknowledged the need and started the process. The protocol specified a freeze on halon consumption. CFCs had to be capped at 1986 levels beginning in 1992, and cut to half those levels by the end of the century. The protocol contained special provisions allowing less-developed countries to come closer to the industrialized nations in their use of refrigeration and cooling equipment.

As scientists learned more about the ozone hole, the world recognized the need for stronger medicine. A 1990 London conference amended the Montreal Protocol. The strengthened protocol called for cutbacks to 50 percent of 1986 production in 1995. Production would be cut to 15 percent of 1986 levels in 1997 and completely stopped in the year 2000.

After the NASA press conference spurred fears that another ozone hole would develop in the Arctic, President Bush speeded up the phaseout of CFCs. According to this new schedule, production of CFCs will stop at the end of 1995, "with limited exceptions for servicing certain existing equipment." Those exceptions might allow as much as 15 percent of 1986 production. The United States now produces less than half the CFCs it did in 1986. Most of the reduction has come from wiser and more conservative use, not from replacing CFCs.

Saying "eliminate CFCs and halons" is easier than deciding how to accomplish the objective. Cutting back on production without eliminating it causes problems. The demand is still out there. If companies make half as much, more people want to buy more of the product than the companies can sell. The price goes up. This means companies can make more money by making less product. Antitrust laws normally prohibit such actions.

Those excess profits can be justified in two ways and rationalized in at least one way. Regulations mandate that companies must get out of the CFC business. They can regard the extra profits as the price paid for their cooperation.

The other justification is better, but harder to understand. Society would like to gain as much as possible from the remaining production. According to economics, the most valuable usage is the one that gives the greatest economic return—and the user getting that return should be glad to pay the most for the product.

The rationalization argues that companies must decide which existing customers receive some product when there is too little to go around. Every customer will describe himself as a faithful customer who really needs the product. The company can either make an arbitrary, perhaps unfair, allocation, or let price increases deter users with less-pressing needs. This is almost, but not quite, the same as the economic argument.

CFC decisions are simple compared to halons. Halons are used in fire extinguishers. They save lives. Some other form of

fire extinguisher might do for the blaze down the hallway, but halons are used in special places—around the engines in jet passenger planes, for example. Because of their special properties, halon extinguishers can be used without putting equipment out of service. How does society decide between its ozone and a jetliner full of passengers that develops an engine fire at 30,000 feet?

Figure 6 shows how society used CFCs in 1986. Some changes have occurred, but these uses yielded our initial cutbacks. CFCs are cheap to manufacture, but few users select CFCs based on price. Users select them because they are so effective. Perhaps the low cost meant that customers used CFCs wastefully. Cutting out waste has already led to major reductions in use.

Little controversy remains about the CFC replacements. The replacements are similar to CFCs, but they include hydrogen atoms that are hooked into the molecule. Hydrogen makes the molecules more fragile, so they break up in the lower atmosphere—where the chlorine does no damage. Replacements fall into two categories: hydrochlorofluorocarbons (HCFCs) and hydrofluorocarbons (HFCs). The HCFCs contain chlorine, and are only a stopgap measure. They will be phased out sometime after the turn of the century. HFCs contain no chlorine. They provide a long-term alternative to releasing ozone-destroying chlorine into the stratosphere.

Everything has a bad side. CFC replacements are more expensive, because it costs money to hook hydrogen atoms into molecules. They are less efficient as cooling agents. Some replacements, unlike CFCs, may be flammable.

ULTRAVIOLET IS NOT INCREASING

Most people believe that ultraviolet exposure must be increasing. After all, the

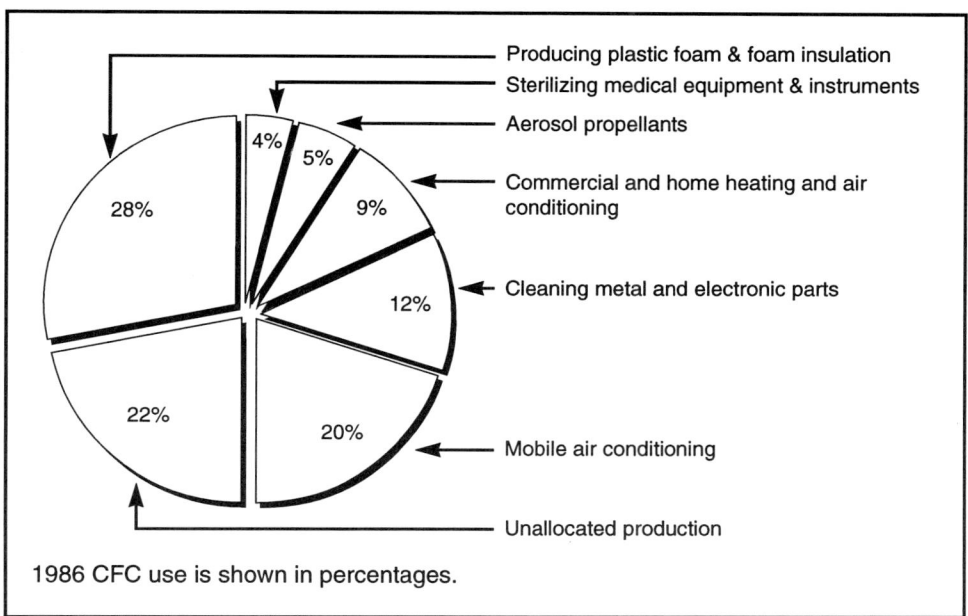

Producing plastic foam & foam insulation
Sterilizing medical equipment & instruments
Aerosol propellants
Commercial and home heating and air conditioning
Cleaning metal and electronic parts
Mobile air conditioning
Unallocated production

1986 CFC use is shown in percentages.

FIGURE 6   *CFCs have many uses besides air conditioning. The percentage used for different tasks in 1986 came from the Environmental Protection Agency. With restrictions on CFCs, the mix of uses is changing rapidly.*

main worry about diminished ozone is the loss of the ultraviolet protection it provides, and the primary harm ultraviolet does to people is cause skin cancers. Thus, increased skin cancers must mean more ultraviolet is getting through.

It may be getting through at the top of the Alps, but it is not increasing where most people live.

Back in the 1970s, the National Cancer Institute, the National Oceanographic and Atmospheric Agency, and Temple University set up a cooperative program to measure increases in UV-B, the ultraviolet that causes cancers. The cooperative set up measuring stations at eight locations across the country—mainly at airports, since that is where weather stations are often located. To everyone's surprise, years worth of data from these stations show no increase in UV-B. In fact, the results indicate a decrease.

When measurements fail to indicate what scientists expected, the measuring system is often attacked. These measurements employed Robertson-Berger photosensitive meters. Many people have recently pointed to shortcomings in the meters. The shortcomings exist, but this does not mean the meters were a bad choice. UV-B is difficult to measure, and every measuring system for UV-B has its shortcomings. A fair assessment regarding the measurements would be this: Shortcomings of the instruments and difficulties in measuring ultraviolet cast some doubt on exact numbers from the experiments. However, the consistency between locations over time indicates that UV-B has in fact decreased.

Skin cancers require some 20 years to appear after the skin is exposed to ultraviolet. Any increases in UV-B resulting from recent decreases in ozone would not yet produce cancers. The increased skin cancers probably come from the way people have behaved in the sun—staying out longer, wearing less clothing, and striving for quick tans.

Tanning salons are a thriving industry, but no one seems to connect the salons with skin cancer. One dermatologist was asked about harm that might result from tanning under a lamp. "It's too soon to say just what harm might come," he replied. "Ask me again in 20 years, after skins have had time to respond, time enough to produce cancers. Time is the only way to be sure."

Those who object to measuring UV-B with the Robertson-Berger meters must contend with another set of measurements. These measurements offer a cause-and-effect theory: that atmospheric haze has caused a decrease in UV-B. Shaw C. Liu and his team from the NOAA laboratories in Boulder, Colorado, took measurements at rural sites in the United States and Europe. He concluded that ground-level UV-B has decreased between five and 18 percent when compared to pre-industrial times.

Haze particles are mainly sulfates and combustion-related aerosols—solid particles produced by industrial activity and automobiles. No one claims we should be proud of them. They spoil the blue of the sky and decrease visibility, but they also may protect us from skin cancer. Ironically, if we clean up the air, ultraviolet might increase. If the ozone level falls far enough, we may have to keep the air polluted in order to protect ourselves from ultraviolet. Some might go so far as to insist that "pollution is the solution"—a gloomy thought indeed.

Liu's measurements came from rural sites. However, he believes that urban measurements would reveal even lower levels of UV-B. "It's a safe bet that the ultraviolet—from air pollution alone—has been cut down something like 20 to 30 percent from pre-industrial times," he reports.

There is another kind of ultraviolet protection that results from pollution. Polluted air contains ozone. We need ozone in the stratosphere, but we try to get rid of it in the troposphere. Ozone is a primary reason that lungs sting when filled with a deep breath of polluted air. Another irony: Ozone cuts down on ultraviolet. Ozone at lower altitudes may even absorb ultraviolet more effectively than ozone in the stratosphere.

Ozone and air pollution may ruin your lungs, but they can reduce the incidence of skin cancer. Have we to some degree become "dependent" on pollution for our health? And how do problems we have not yet discussed—things like global warming—fit into the equation?

3

The Greenhouse Effect

They should call it the parked-car-on-a-sunny-day effect.

In 1827, French physicist Jean-Baptiste Fourier suggested an analogy in order to explain how air helps keep the world warm. The world, Fourier said, stays warm because air traps heat, as if under a pane of glass. He was lucky enough to be right. He made his analogy before anyone knew about infrared radiation and radiative cooling.

Greenhouses work the same way. They protect plants from freezing by keeping them inside a heated house. Plants need light, so horticulturists construct the houses from a material that lets light through. During the night, the heater keeps the plants warm and unfrozen. During the day, the sun shines in and provides light to allow the plants to grow. Here, a problem develops. The glass lets the sun shine in; it lets in the solar energy. The glass traps the solar heat, just as it trapped the heat trapped during the night. Trapping sunlight means the greenhouse often gets too warm, and that trapping of heat is the greenhouse effect. The way the earth's atmosphere traps heat is an imperfect analogy to the way greenhouses overheat during the daytime.

By 1896, enough was known about infrared radiation and carbon dioxide that a Swedish geochemist named Svante Arrhenius could make a good greenhouse heating prediction for the world. Even then, the large amounts of carbon dioxide released into the air by human industry worried Arrhenius. He was way ahead of his time. According to his estimates, doubling the amount of carbon dioxide in the air would cause the world to warm by 7 to 11°F. That estimate and the recent estimate given by the International Panel on Climate Change predict roughly the same amount of heating. Arrhenius made his estimate without computers, satellites, or the other trappings of modern science. Yet, even without a grant from the National Science Foundation, he got the same answer we get now. However, no one has paid much attention to either answer.

The years around 1896 were a great time for science. During the same year in which Arrhenius estimated greenhouse heating, Henri Becquerel discovered radioactivity. And Hertz verified the existence of radio waves in the late 1890s. In one short period, we arrived at the basis for broadcasting, nuclear weapons, and environmental concerns that still need sorting out.

In truth, Arrhenius was lucky to get an answer close to the current one, and for all we know, both answers may still turn out to be wrong. If so, it will not be because carbon dioxide and other greenhouse gases fail to generate heating as predicted. The error will occur because the world is a complex place, and some other compensating effect may cancel the heat-trapping effect of the greenhouse gases.

After all these thousands of years, you might expect the weather to have settled down and come to equilibrium. Stir a bucket of water; wait a while; the circulation slows and finally stops. In the atmosphere, however, the wind keeps blowing and things never settle down. The atmosphere keeps circulating because it is a heat engine. The sun stirs the bucket by depositing heat. The atmosphere redistributes the heat, then gets rid of it. People seldom think about how the earth gets rid of its heat, but it must—otherwise the earth would get hotter year after year, and at a more rapid pace than predicted by the current global warming calculations.

This chapter takes a broad view of what greenhouse heating is. Rather than asking how the greenhouse gases might affect future temperatures, the broad view asks what is happening now. It examines how the greenhouse effect couples with other interactions and explains the combined effects. Such interactions cause ground level temperatures to cool each night. They also explain why air gets cooler at higher altitudes, and why winds blow in the direction they do. All this is less complicated than it sounds. Understanding a few physical principles can make everything seem simple. The first thing required is a description of the atmosphere, including what it is made of and how it is structured.

THE ATMOSPHERE

Grade school health lessons explain how we breathe in oxygen and breathe out carbon dioxide. As a result, most people think the atmosphere contains little more than the two gases. Air does contain a considerable amount of oxygen, but oxygen is not the most common gas in the air. Nitrogen is the most common: Air is 78 percent nitrogen, only 20 percent oxygen. Air contains only a trace of carbon dioxide—0.033 percent, or 330 molecules of carbon dioxide per each million molecules of air. Such a small amount seems hardly worth mentioning. It gets mentioned because carbon dioxide is so important.

Animals must have oxygen to survive, while plants must have carbon dioxide. Plants and animals exist in a symbiotic relationship. We convert oxygen into the carbon dioxide that plants need; plants recycle our waste product—

carbon dioxide—back into the oxygen we need. The truth is, we get the best of the deal. Plants came along millions of years before animals. They can get by without us, but we cannot get by without them. Still, we talk about the symbiotic relationship as if we were as vital to plants as they are to us. That relationship is why we hear so much about oxygen and carbon dioxide and so little about nitrogen. Another gas that rarely gets mentioned is argon. Air contains more argon than carbon dioxide, but argon plays no part in atmospheric chemistry.

Water is discussed so often because living things need water, just as they need oxygen or carbon dioxide—but water is important for other reasons. The world has a hot water heating system. The tropics do not need heating, but the polar regions surely do. Ocean water gets hot in the tropics, then currents transport it north. Once there, the warm water helps heat the polar regions. Without water circulating from the tropics, the polar regions would be much larger and much colder.

Water drives atmospheric circulation and heating because it evaporates, condenses, and freezes at temperatures regularly found in the atmosphere. More than that, it absorbs a great deal of energy as it evaporates, and releases energy when it condenses or freezes. Cooling a cup of coffee with spoonfuls of ice or ice water offers one example of how much energy ice absorbs when it thaws. One teaspoon of ice cools the coffee as much as 80 teaspoons of ice cold water. Evaporation is even more spectacular. It takes almost seven times as much energy to evaporate water as it does to freeze it.

Letting one teaspoon of water steam off the top of coffee cools it as much as 537 teaspoons of ice water. That is why putting covers on cups or dishes cause them to cool slowly; the cover stops evaporation.

Oxygen, nitrogen, and carbon dioxide make up the same fraction of the air throughout the lowest 30 miles of our atmosphere. At higher altitudes, the air is too thin to be important in greenhouse cooling. Some other atmospheric gases vary in concentration from place to place, due to sources that generate the gases and sinks that absorb them, or vary with altitude due to chemical interactions.

Methane, sometimes called swamp gas, varies in concentration because local sources release it. Consider, for example, the air downwind of a livestock feedlot or a sewage treatment plant. That air contains much methane in addition to the other gases, giving the air its strong odor. However, methane itself is odorless. Methane is also converted to water vapor in the middle atmosphere. That water vapor in the middle atmosphere freezes, creating some of the ice particles that help the chlorine from CFCs create the ozone hole.

Air temperatures fall at an almost constant rate as altitude increases. Air cools about 2°F per 1,000 feet of altitude. Such cooling cannot continue forever, and in fact it stops at about 50,000 feet—at the atmospheric layer called the tropopause. At the tropopause, temperatures stop getting colder with altitude. They indeed pause, and temperatures remain stable for the next few thousand feet. After the pause, temperatures increase.

The lower atmosphere, where temperatures decrease with altitude, is called the troposphere. The region where temperatures increase with altitude is called the stratosphere. Infrared cooling causes temperatures to get colder at higher altitudes in the troposphere, but not in the way a reasonable person might expect. Absorption of solar energy, mainly by ozone, causes the warming at higher altitudes in the stratosphere.

Atmospheric pressure—i.e., barometric pressure—is the weight of all the air overhead. People never think of air as having weight, since it floats around rather than falling to the ground. Science classes prove air has weight by weighing a basketball with and without air. In fact, air weighs about two pounds per cubic yard. The weight of air causes the high and low barometric pressures featured on weather reports.

At sea level, barometric pressure is 14.7 pounds per square inch. Consider what that square inch means. Let's say someone erects a long tube with a one-square-inch cross section, extends it from the ground to the top of the atmosphere, closes the top end, then cools the tube until all the air in the tube turns to liquid and freezes. The frozen material from this square inch of atmosphere would create a bar some tens of feet long that would weigh 14.7 pounds. This means that, at sea level, there are 14.7 pounds of air over every square inch of this planet. Because only 0.033 percent of that air is carbon dioxide, amounting to less than one-tenth of an ounce, it is fairly easy to double the amount of carbon dioxide in the atmosphere. The air contains even less methane.

As one goes higher in the atmosphere, there is less air above. Less air means less weight, so going higher means decreased barometric pressure.

Here is what happens when pressures fall off. Each time an airplane gains 18,000 feet in altitude, it passes through half the weight of the air above it. Stated another way, each time altitudes increase 18,000 feet, the amount of air above—hence, the barometric pressure—drops to one-half of the previous value. At an altitude of 18,000 feet, freezing all the air in a square inch above you would yield only 14.7 divided by 2, or 7.35 pounds of frozen air. Ascend another 18,000 feet, to 36,000 feet, and the amount of air above you drops by half once again, to 3.67 pounds per square inch.

Temperatures also drop at a constant rate with altitude. At 18,000 feet, the temperature would be 36°F cooler than at the ground, and at 36,000 feet it is ould be 72°F cooler. We can arrive at a sense of how fragile life on earth really is by considering the air at 18,000 feet—2,000 feet below the top of Mount McKinley. Even at such altitudes, low barometric pressure and freezing temperatures make it impossible for trees, most animals, and all but a few plants to live. Many people can survive without oxygen masks at this altitude, but some would pass out without additional oxygen. At the 29,000-foot altitude of Mount Everest, essentially nothing can live without a life support system. Life is so fragile that nothing can live on some parts of our world. And the way in which we are now living may soon make it impossible for life to survive on other parts of the planet.

Atmospheric pressure at two different locations can be different, even when the locations are at equal altitudes. That is why weather maps indicate highs and lows. The weight of air is greater above the place with higher pressure. One way this difference comes about is through temperature. Temperatures above the two locations may differ. If you heat air, it wants to expand. The only way to heat air without expansion is to confine the air and let its pressure increase instead of its volume.

High pressure regions, obviously, are subject to high barometric pressure. Naturally, pressure at any given altitude attempts to equalize itself. Everyone expects higher pressure air to expand into the lower pressure region—to fill the vacuum. The attempt to equalize pressure after temperatures change is what makes the winds blow and keeps air on the move. As long as some places get hotter than others, the wind will keep blowing.

A television weatherman in Denver used to say, "Wind is air in a hurry." He never got around to explaining why the wind was not flowing directly away from the high pressure region, as would seem likely.

Getting from the high pressure zone to the lower pressure zone is not as simple as it sounds. Because the world is rotating and angular momentum remains unchanged as the air moves, strange things happen. In the northern hemisphere, the earth's rotation causes the wind to deflect to the right of the direction in which it is moving. In the southern hemisphere, the deflection is to the left. This effect is called Coriolis deflection. It makes air take a round-about path between high and low pressure regions. If air constantly keeps moving to the right of the direction it is traveling, it ends up traveling in a circle. So, when air tries to move from high pressure to low pressure regions, it spirals inward. This circuitous path takes longer, but eventually pressure between the two locations becomes equalized.

Why is wind deflected to the right in the northern hemisphere and to the left in the southern hemisphere? The person in the northern hemisphere looks up and says the world spins counterclockwise. People in the southern hemisphere are effectively standing upside down with respect to those in the northern hemisphere. Their "upside-down" view causes them to see the world as spinning clockwise. Both sets of observers are correct from their own vantage point. Wind tries to compensate for the rotation, so it rotates in the opposite direction. That means wind deflects to the right in the north, to the left in the south, and not at all along the equator.

The lack of deflection at the equator limits the transfer of air between the northern and southern hemispheres. Air moves *along* the equator instead of across it. One result is that carbon dioxide concentrations tend to be higher in the northern hemisphere, which produces more carbon dioxide than the southern hemisphere, mainly because it contains more land area and more industry. However, the carbon dioxide from the northern hemisphere eventually gets into the southern hemisphere; the process just takes a while, and meanwhile the north has accumulated even more carbon dioxide.

COMMON RADIATIVE EXPERIENCES

All the theory and experiments discussed here explain the radiative heating and cooling of the world. The term "radiative" has nothing to do with radioactivity; it describes the way objects emit heat into their surroundings, just as a hot stove emits heat. Theories may seem somewhat abstract, so it is useful to look at common experiences in order to demonstrate that the theory is valid.

Few of us have ever gone into a greenhouse on a sunny day to see if it is warmer inside than outside. Even if we did, greenhouses have fans that pull in the cooler outside air. We hear about something called the greenhouse effect; but this effect might better be expressed by comparing it to something everyone experiences—getting into a parked car on a sunny summer day.

Car windows work in the same manner as the greenhouse windows. Visible light can get through them, but the windows do not transmit the infrared. The windows prevent any air in the car—or emission from the seats or the dashboard—from escaping. There is no way for the car to cool itself except for the small amount of heat conducted out through the glass, and glass is a poor conductor of heat. The result? Greenhouse heating in your car.

The automobile can be used to verify the greenhouse effect in another way. The reason greenhouse heating causes the surface to warm is this: Putting more absorption in the atmosphere blocks surface cooling. Consider what happens to a parked car on a cold night.

On a really cold night, frost develops on the windshield, and on the rear and side windows of a car. On a moderately cold night, the windshield and rear window of the car may frost up, but no frost appears on the side windows. The lack of frost on the side windows actually verifies the greenhouse effect.

Think about the shape of a car. The windshield is sloped, which means it faces upward. Side windows are vertical, so they face outward. Facing up means "looking" through less atmosphere than facing out. The difference in slope explains the difference in frost formation. At night, the side windows cannot "see" much of the sky directly above them because they face to the side.

What does this have to do with the greenhouse effect? Looking out through the atmosphere at an angle means looking through more atmosphere than when looking vertically. That is why the horizon always looks more polluted than the sky above. The side windows have to look through a greater thickness of *air.* Looking through more air means looking through more carbon dioxide, water, and all the other greenhouse gases. Windows that have to look out near the horizon look through the most greenhouse gases. Therefore, the vertical side windows have to look out through more carbon dioxide. Looking though a longer atmospheric path causes these windows to cool more slowly, and to develop less frost on cold nights. They cool less because the thicker atmosphere emits more radiation to warm them.

We'll cover the radiative effects of clouds in another chapter, but one effect helps explain nighttime cooling. The coldest winter nights occur when skies are clear. On nights when clouds cover the sky, little radiative cooling occurs. The clouds turn cooling off by capturing the radiation from the ground. After capturing this heat, the clouds reradiate it, and half of it heads back downward. This means the ground recaptures almost half the radiation it emits. Winter clouds are usually low, so they are at nearly the same temperature as the ground. The clouds thus effectively block radiative cooling, and temperatures hold at about the same level they were at when the sun went down.

During cold, clear weather, the air can hold less water vapor. This is why cold windows build up condensation—the air next to the window cools and deposits its moisture on the window. Normally, water vapor serves as a greenhouse gas and limits nighttime cooling. On cold nights, the reduced concentration of water vapor lets temperatures drop more rapidly than on warmer nights, when the air can carry more water vapor. So clear winter nights can cool more rapidly and become frigid by morning.

Warmer air does not necessarily contain more water vapor. It has the *potential* to hold more water vapor than colder air.

Deserts are another good example of how important water can be as a greenhouse gas. The standard example illustrating water's function as a greenhouse gas involves desert nights. Because deserts usually have few clouds, all the incoming sunlight can reach the ground. Deserts contain dry air almost devoid of water vapor. The limited amount of water vapor absorbs some sunshine, but not too much. Recall that greenhouse gases absorb thermal infrared light, which means they reradiate heat back down to the surface, which keeps the surface warmer because the ground's radiative cooling is counteracted by radiation bouncing back from the sky.

Come nighttime in the desert, the air has almost no water vapor for trapping heat. As a result the desert, where the sun heats the surface to egg-frying temperatures during the day, gets quite cold at night—sometimes cold enough to freeze. These cold temperatures illustrate water vapor's effectiveness as a greenhouse gas. To see a comparison between a humid and a dry atmosphere, notice the daily minimum and maximum temperatures in, say, humid Houston and dry Phoenix. Phoenix experiences much greater daily temperature swings than Houston. Less water vapor in the air means more cooling at night, so the temperature variations are greater.

RADIATIVE HEAT EXCHANGE

If you turn on the burner of an electric stove, it starts to heat. As it heats, it changes from emitting no glow, to a very dull red glow, to a bright red, and finally to a reddish-white color. Even before the burner starts to glow, it may be too hot to touch.

As an object gets hotter, it radiates at shorter and shorter wavelengths. The dull red corresponds to radiation so long in wavelength that the human eye can hardly see it. As the burner gets hotter, its color shifts toward the blue, shorter wavelengths. Measure the color, and you can arrive at the burner's temperature. Scientists measure the temperature of the sun by using this method—essentially, they measure its color. The color of the sun results from emission of a broad range of wavelengths, not just a single wavelength.

Even though a body emits some wavelengths short enough for the eye to see, it may also emit longer wavelengths, such as infrared, that are not visible. Hot bodies emit more short wavelengths, but longer wavelength radiation also increases. Emission at the longer wavelengths, however, does not increase as rapidly as emission at the shorter wavelengths.

When something—a rock, for example—is hotter than its surroundings, it cools by radiating heat. The hot burner does the same thing. The cooler surroundings absorb its radiation and get warmer. Even in a vacuum, everything would eventually become the same temperature by exchanging such heat radiation—after all, heat from the sun comes to earth through the vacuum of outer space.

Rocks radiate all the time. The hotter the rock, the more it radiates. Since even vacuums transmit radiation, every object, including the rock, constantly receives radiation from its surroundings. The hotter the surroundings, the more radiation the rock receives.

This is a self-adjusting process that does not require that the rock decide whether to radiate. If it is hotter than its surroundings, it radiates more than it receives, and cools off. If it is colder than its neighbors, it absorbs more radiation from the surroundings than it radiates. It then continues warming until it reaches the same temperature as its neighbors. Once at this temperature, it receives and radiates the same amount, which means it stops warming.

There is one other point to make about emission and absorption by rocks. Suppose there are two rocks—one that absorbs almost all the radiation striking it, and one that absorbs only part of the radiation. When the two rocks are sitting next to each other, they come to the same temperature. If one absorbs a larger fraction of the radiation hitting it than the other, it must also emit more. Otherwise, the more absorbent rock would get the hottest. The rule is this: high absorption means high emission.

The sun covers only one small spot in the sky around the earth. Most of what surrounds us is the cold nothing of outer space. Earth will never get as hot as the sun, because it is not completely surrounded by the sun. Instead, the world warms until the amount of heat it radiates in all directions cancels out the amount of radiation it receives from the sun. Earth radiates at all angles, but only receives radiation from the small angle covered by the sun. This lets the world balance its heat budget while remaining at life-sustaining temperatures.

We have covered almost all of the rules we need to understand in order to comprehend the greenhouse effect. One rule, concerning reflection and absorption, bears repeating. An object that reflects a lot of radiation at one wavelength emits only a little radiation at that wavelength. Likewise, an object's emission increases as it becomes absorbent. An object that emits a lot of radiation had better absorb a lot, or it will become colder than everything around it—and objects never stay colder than their surroundings.

It helps to point out that how transparent an object happens to be in the visible realm has nothing to do with how transparent or absorbent it is in the infrared. In the visible, glass is quite clear; in the infrared, glass is opaque, or black. Black, in this context, means highly absorbent. A black object absorbs and emits more than any other object at the same temperature. Restaurants often wrap baked potatoes in foil before serving them, because foil reflects infrared radiation instead of absorbing it. The low absorption of foil means it also emits infrared poorly, so it loses little heat through radiation, which helps keep the potato warm.

Here is an example of how radiative heating works. Imagine moving a window around to find out the importance of its location in the atmosphere. Or, instead of a window, use a layer of gas that behaves much the same as glass. The gas is transparent to visible light but only semitransparent to infrared radiation. It transmits, say, three-fourths of the infrared and absorbs the other one-fourth. Objects that absorb poorly emit just as poorly. Since the gas absorbs just 25 percent, it is not fully black in the infrared, and emits less than it would if it were totally black, or opaque.

How much the layer of gas emits depends on its temperature. Here is where moving the layer up and down in the atmosphere becomes important. As it moves up and down, its temperature changes. Figure 1 shows the different streams of radiation to and from the layer.

The layer absorbs a quarter of the radiation from the ground, but that is of little interest here. From the ground's viewpoint, it rid itself of all the radiation

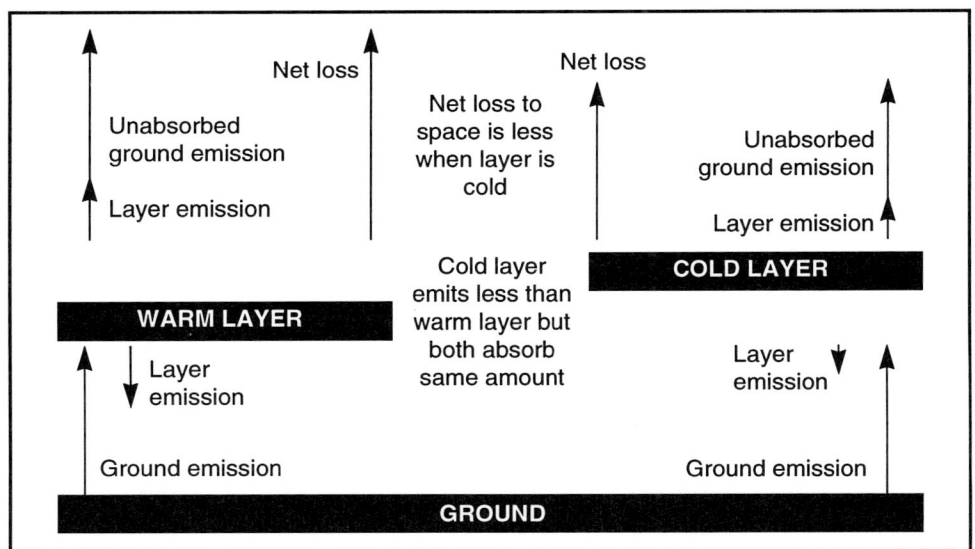

FIGURE 1 *High, cold layers of gases cut down more on radiation losses than low, warm layers. Both layers shown here absorb the same amount of radiation from the ground. The warm layer emits more radiation to space and has a greater net cooling effect for the earth. Exchange of radiation between the layer and the ground partially determines the layer's temperature.*

it emitted. Where that radiation goes may be important in deciding the temperature of the atmosphere, but the ground is unaffected by where the radiation ends up. What does affect the ground is the radiation coming *from* the layer. The layer absorbs one-fourth of the infrared striking it, so it only emits the same fraction.

This level of detail about the layer is important in understanding greenhouse heating. How much heating a gas causes depends upon how high—in other words, how cold—the gas is. This works for clouds as well as gases.

No matter where the layer is, it emits one-fourth as much as a black body would emit at the same temperature. If the layer is near the ground, the air is warmer, so it emits more and warms the ground to a greater degree than it would if it were high and cold. When it is cold, the layer sends less radiation to the ground, warming the ground less. Moving the layer has no effect on what the ground emits. Unless the ground heats or cools, it always emits the same amount of radiation. The ground's net loss is greater with a high, cold layer than with a low, warm one. The high, cold layer emits less radiation into space, too, which cuts the earth's net loss.

The preceding paragraphs explain everything that is needed to understand greenhouse heating. If you go through them again, you may understand more about greenhouse heating than many science reporters. Having come through it the long way, the whole greenhouse principle can now be stated in three sentences:

Layers of absorbent gas above the ground cut down on energy losses from the ground.

Low layers cut losses more than higher layers, and they also diminish the effects of the higher layers.

If we had a choice of where to put greenhouse gases, placing them high would cause less warming.

What does all this mean to the average temperature of the world? The sun radiates like a black body at 11,000°F. But the world is a long way from the sun, and moving back from a heater cuts down on the heat received. Also, the sun shines on only one side of the earth at a time, but the earth emits radiation from every square inch of its surface at all times. The world has a heat budget. If the earth emits less than it absorbs from the sun, it gains energy and warms. As the earth warms, the amount of heat it radiates increases. This process determines the world's average temperature.

Figure 2 illustrates how much energy the earth receives from the sun at each wavelength, and how much it emits back into space. If the sun emitted as a black body, this figure would be just right. In reality, it does not, but the approximation is fairly accurate. The horizontal axis in the figure shows the wavelength of the radiation, and the vertical axis shows how much radiation is received at this wavelength. Wavelengths are given in micrometers (one micrometer is 0.000254 inch).

Earth is cooler than the sun, which means its emissions occur at longer wavelengths than emissions from the sun. At wavelengths of less than four micrometers, the sun provides more heat than the earth emits. At wavelengths of longer than four micrometers, the earth emits more radiation than it receives. People who work with infrared call the wavelength region beyond four micrometers the thermal infrared.

Knowing how much radiation the earth gets from the sun makes calculating the effective temperature of the earth an easy proposition. The effective temperature is the uniform temperature needed for the earth to emit precisely the same amount of energy it receives from the sun. The real world is more complex, of course, but for the moment it is useful to simplify things. To help simplify the equation, think of the atmosphere here as a single layer of gas.

Here is where greenhouse heating comes in. Suppose the layer of gas is black in the infrared. Then none of the radiation from the ground gets through the layer. Radiation from the layer would be the only radiation escaping the earth, because radiation from the ground cannot penetrate the gas layer. The earth would finally warm up and become as hot as the gas layer. And that is the why greenhouse heating causes global warming.

Atmospheric absorption blocks radiation from the ground, and the earth's effective emission temperature is lower than the ground temperature. The atmosphere is largely transparent to the radiation from the sun, letting most of it reach the ground. Only after the ground warms up and then warms the atmosphere can the incoming and outgoing energy reach a balance.

In truth, air does absorb some short wavelengths from the sun. The air is also

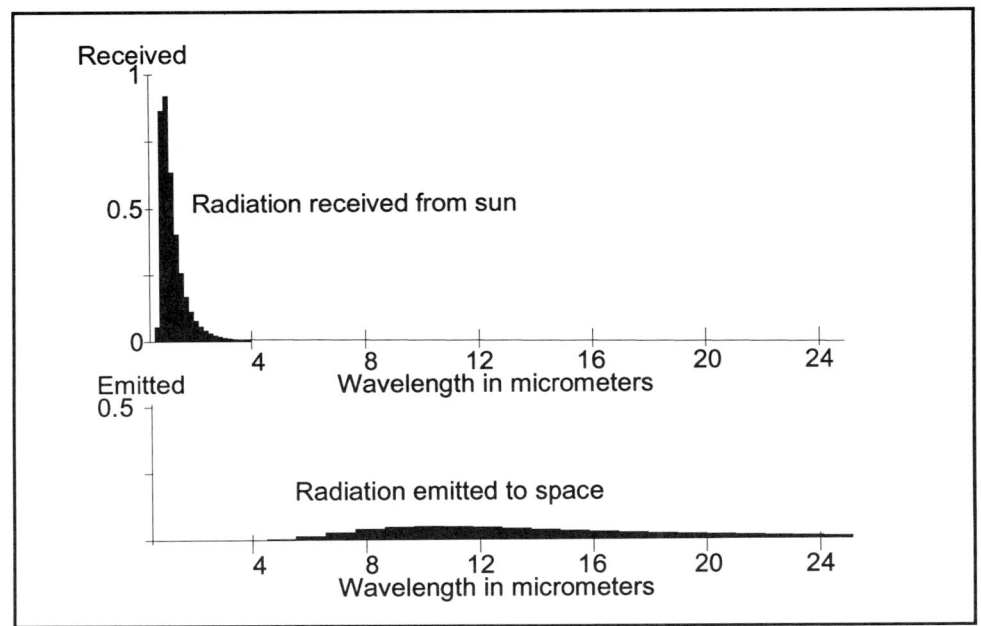

FIGURE 2 *The two graphs show the energy received from the sun and emitted by the earth on a scale that balances the incoming and outgoing gains and losses. The incoming solar radiation is intense at shorter wavelengths, while the infrared emission is less intense, but is spread over a wider range of wavelengths.*

partially transparent at some infrared wavelengths. It is time to consider the absorption spectra of the air and the sun as they really are, instead of simplifying them.

ABSORPTION SPECTRA

Some gases absorb at one wavelength, some at another. Some do not absorb at all. Gases are about as individualistic as people. Some gases seem to do nothing. Take nitrogen for example. It is 78 percent of the air, but absorbs neither visible nor infrared wavelengths. Oxygen, another 20 percent of the air, is hardly more interesting. It absorbs only a small amount. This means that 98 percent of the molecules in air absorb next to nothing. The other two percent of the gases do all of the absorption and emission. It seems unjust to refer to the active two percent as trace gases, but that is what atmospheric scientists call them. If so little of the atmosphere can accomplish so much, it makes sense that the atmosphere can be highly sensitive to a small amount of pollution.

Most atmospheric gases exist as molecules, not as atoms. Nitrogen and oxygen molecules, respectively, consist of two atoms of nitrogen or oxygen joined together. Ozone, as we've seen, is three atoms of oxygen. Carbon dioxide, as the name implies, consists of one carbon atom and two oxygen atoms. Everyone recognizes water as H_2O—two hydrogen atoms and one oxygen atom. Methane, the primary gas in natural gas, is a carbon atom surrounded by four hydrogen atoms. When methane burns, the carbon joins with atmospheric oxygen to form carbon dioxide, and the hydrogen links with oxygen to form water. When burning methane, more heat comes from the formation of water out of hydrogen and oxygen atoms than from the formation of carbon dioxide out of carbon and oxygen.

Elementary science teachers often say: "Burning occurs when carbon and oxygen combine to form carbon dioxide." Burning does produce carbon dioxide, but all the hydrogen in organic matter has to go somewhere. The burning converts it to water, just as we've seen with methane. Perhaps water is left out of educational explanations of burning because we think of water as something that puts out fires. Some engineers advocate using hydrogen as fuel. It would burn and release only water, producing no carbon dioxide at all.

Whether a molecule absorbs or not depends on how its electrical charge is spread within the molecule. There must be some sort of asymmetry if the molecule is to absorb. Nitrogen has no "charge asymmetry," so it does not absorb. Oxygen has only a mild asymmetry, caused by the way one of its electrons spins, and so it only absorbs weakly. Water, on the other hand, is highly asymmetric. It absorbs strongly at several wavelengths. The water molecule has the shape of a sharp triangle with the oxygen at one tip. The oxygen appears negative, and the hydrogens at the other tips of the triangle are positive, creating asymmetry.

If you ever want to hear something really confusing, ask a physicist about absorption. First, she or he will talk about light as an electromagnetic wave that interacts with the charge distribution in molecules. As the discussion develops, the physicist will explain that the light is not really a wave. Instead, it consists of particles called photons. Ask for clarification concerning whether light is a wave or a group of particles, and the opposite of clarification will follow.

If translated into ordinary human language, the clarification goes something like this: Some scientific problems demand that light be considered a wave. Using a lens to focus photons instead of light waves, for example, causes real difficulties. Other scientific problems, such as studying absorption by molecules, are solvable only if light is treated as a particle. Whether light is a particle or a wave, according to the physicist, is not a fair question. Sometimes it is considered one, sometimes another. The physicist finishes by claiming that this ambiguity is acceptable, because light is actually neither one. At this point, it is a good idea to flee from the discussion.

Figure 3 tells the full story of how and at which wavelengths air molecules absorb. The figure is broken into sections—one section for each gas and one at the bottom for absorption by all the gases combined. Visible wavelengths occupy only a small fraction of the figure, ranging from 0.4 to 0.75 micrometers. Wavelengths shorter than 0.4 micrometers are in the ultraviolet, and those longer

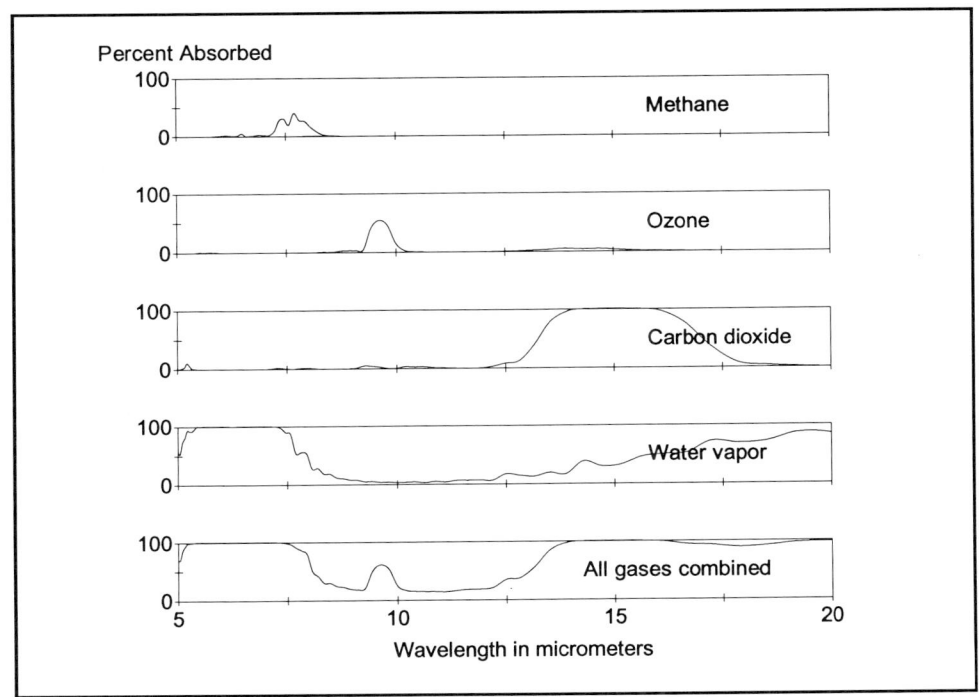

FIGURE 3 *How important a gas is in creating global warming depends on how much it absorbs and how close that figure is to the peak of the emission curves shown in Figure 2. Water absorbs some at almost all wavelengths, which makes it quite important. CFCs are not shown, but they absorb near ten micrometers, the peak of the emission curve, where the atmosphere is quite transparent.*

than 0.75 micrometers are in the infrared. Humans see neither ultraviolet nor infrared wavelengths. Figure 2 indicated that much of the sun's energy is in the infrared, but at wavelengths shorter than the thermal infrared, where atmospheric emission is important.

Water is certainly the busiest of the absorbers. It absorbs at long, short, and intermediate wavelengths. Figure 3 justifies saying that water is the most important of the greenhouse gases. Carbon dioxide, which receives the most attention as a greenhouse gas, does not seem so impressive as an absorber.

The figure shows the percent of radiation that's absorbed when moving straight up through the full atmosphere. To understand why carbon dioxide is so important, refer again to Figure 2—the black body curves. The absorption by carbon dioxide occurs near the peak of the atmospheric black body curve. Carbon dioxide blocks an important section of the thermal infrared. Carbon dioxide absorption occurs near the area where the earth's emissions are greatest. This means CO_2 is more important than one would guess by looking at Figure 3.

Ozone has three claims to fame. First, it absorbs in both the ultraviolet and the infrared. Second, its infrared absorption takes place near the peak of the atmospheric black body curve, just like carbon dioxide's. The third characteristic cannot be seen in the figure. It involves the placement of ozone in the atmosphere— at high altitudes. Ozone does most of its absorption at high altitudes, where ozone concentrations are highest. Because it is both high and at a low temperature, ozone absorbs a great deal of the radia-

tion from the warm ground. Also, ozone emits little radiation to replace its absorption, because it is cold. In turn, this really diminishes the amount of radiation escaping from the earth. No one talks about ozone as a greenhouse gas. If the ozone goes, global warming will slow down, but let us hope that something else resolves the global warming problem.

Figure 1 represents what the atmospheric emission would be if the atmosphere were opaque at all wavelengths. Figure 3 shows just how opaque the atmosphere is at different wavelengths (zero transmission is black or opaque, 100 percent is transparent). The atmosphere is not all that black in the infrared, but the ground is. In regions where the atmosphere is semi-transparent and absorbs little, radiation from the ground can get through the atmosphere, which allows the ground to cool.

Putting more gases like carbon dioxide or methane into the atmosphere— and thus reducing the atmosphere's transmission— blocks radiation from escaping. Less escape means less cooling, so the ground stays hotter. Because it is warmer than the air, the ground heats the air. Ultimately, absorbent gases do the same thing as windows in a greenhouse. They let heat in, but they also keep it from escaping.

The sun does not radiate like a black body. It has its own atmosphere, which means it has its own problems. At some wavelengths the sun's atmosphere absorbs radiation from deep within the sun, because the sun's outer atmosphere is cooler, just like the earth's. In other regions, the sun's atmosphere is more transparent and lets emission get through from deep within the sun. The combination of absorption and emission makes radiation from the sun a more complex mixture than the idealization shown in Figure 2.

Solar gases have absorption spectra just like our atmospheric gases. One example of solar absorption spectra involves helium. Helium is such a light molecule that essentially all the helium in the earth's atmosphere escaped through the top of the atmosphere, boiled away into outer space. Since there is no helium in the atmosphere and helium takes part in no chemical reactions that might keep it locked up in the earth, it was a hard element to discover. Strange as it sounds, helium was first discovered on the sun. In looking at the solar spectrum, lines could be seen that

corresponded to nothing known on earth. The element responsible for those lines seemed to exist only on the sun (Helios), which accounts for its name.

Figure 4 shows the real emission of the sun, not the idealization. This is the solar radiation at the top of the atmosphere, before the air absorbs any of it. Beneath this curve is the solar radiation reaching the ground. Atmospheric gases absorb part of the solar radiation. Molecules and dust particles in the air (aerosols) scatter part of the radiation back into space. At shorter wavelengths, scattering has more effect than absorption.

Molecules do more scattering at the shorter wavelengths than at the long ones. This explains why the sky is blue. Blue is at the short wavelength end of the visible spectrum, and the blue light from the sun that would normally shine

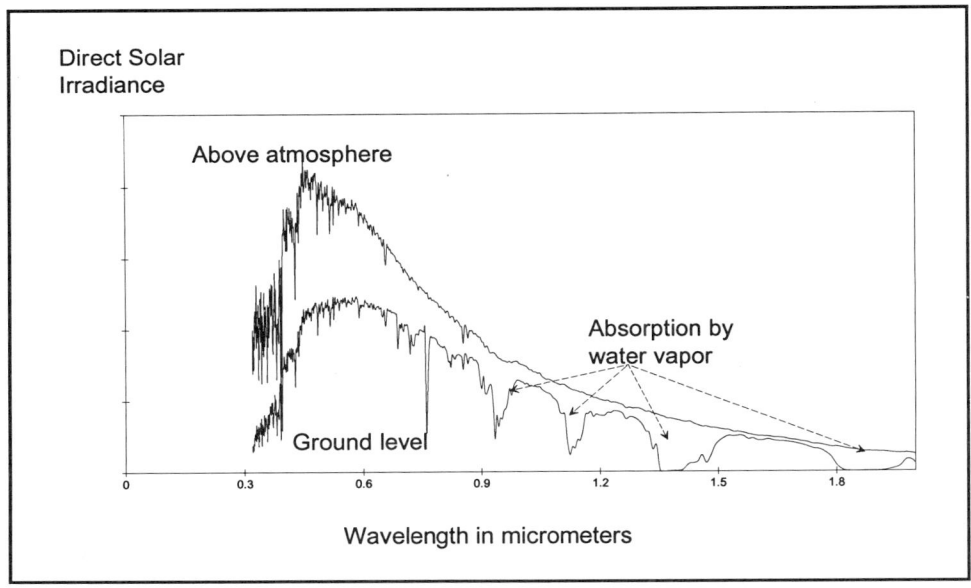

FIGURE 4 *Much more direct solar radiation reaches the top of the atmosphere than reaches the ground. Only a few wavelengths are absorbed. Most of the decrease is due to scattering, which is greater at the shorter wavelengths. Besides absorption by water, ozone also absorbs near 0.3 microns. This is the ultraviolet protection that ozone provides.*

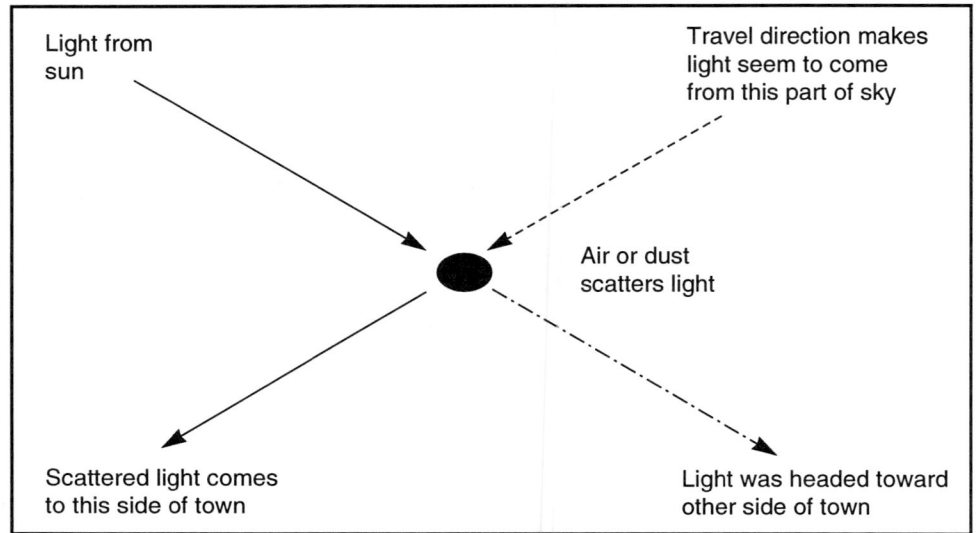

Light from sun

Travel direction makes light seem to come from this part of sky

Air or dust scatters light

Scattered light comes to this side of town

Light was headed toward other side of town

FIGURE 5 *The sky is bright because air and small particles in the air scatter incoming sunlight. Our eyes tell us that light comes from the direction in which we see brightness. Because scattering is greater at the blue wavelengths, the sky looks blue.*

down on us gets scattered and diverted. As Figure 5 shows, scattered light comes from a different direction than the direction of the sun. The light seems instead to come from the sky, and since most of the scattered light is blue, the sky appears blue. In the mountains or on a high-flying jetliner, there is less air above to scatter light. Less scattered light means the sky is not so bright and looks a deeper, darker blue.

On the other hand, dust particles scatter more red and green light than molecules do. Instead of making the sky appear blue, such scattering makes it appear lighter-colored, giving it a dirty, milky appearance.

While the sun sets, its rays must pass through a much longer atmospheric path than is the case when they are coming almost straight down. The path is so long, in fact, that all the blue and green light

scatters and re-scatters, until no blue or green remains. Air scatters very little red, but clouds and dust scatter red quite well. The clouds and dust make the sky appear red at sunrise and sunset. All the other colors have been scattered or absorbed, and red is the only color left to see.

DAY-NIGHT TEMPERATURES

Days are warm and nights are cold. The explanation normally offered is that the sun shines during the day and heats the air. However, the point was made earlier that air absorbs little sunshine. It sounds like one of these statements must be wrong, but in fact both are correct. Air absorbs little heat directly from the sun, yet the sun does heat the air. More accurately, the sun heats the ground, and the ground then heats the air. So the sun does indirectly cause warmer air. This short explanation is only the tip of some atmospheric processes that are vital in determining atmospheric temperature profiles.

First, there is the matter of how much the sun heats the ground—a complex question. Rocks in the desert get so hot you can fry eggs on them, but deserts also reflect more sunlight than green fields. Why, then, do deserts get hotter than green fields? Since air is heated by the surface, not by the sun, the air can get no hotter than the surface. So desert air gets hotter than air above grassy meadows—because the desert surface is hotter.

Why isn't the surface of the meadow as hot as the desert surface? Two things are happening. Part of the sunlight reaching the meadow is used for something besides heating. It causes plants in green fields to grow. That growth converts the energy in sunlight into the chemical bonds formed when photosynthesis takes place and vegetation grows. If you burn the plants, the heat from burning releases the solar energy that the plants absorbed when growing. If you let the plants decay, they release the same amount of energy, although at a slower rate.

Because plants grow, they store some solar energy through photosynthesis. However, photosynthesis uses solar energy so inefficiently that the EPA should consider requiring energy-efficiency labels on house plants. Most plants convert less than one percent of the sunlight striking them into plant growth. Nonetheless, plant growth uses too little energy to explain why the green meadow is not as hot as the desert.

The relationship between plants and temperature is important to global warming because it provides a feedback mechanism. How much global warming we get depends on how plants do in the warmer climate. If plants do not grow well, the world will become more desert-like and even hotter. If plants thrive, their success means that the world will not get so warm. Since the cooling mechanism is so important, and since that mechanism is not photosynthesis, here is the explanation:

Plants have an evaporative cooler. Their leaves are quite dark, reflecting only a small amount of radiation. Deserts may reflect 30 or 40 percent of the sunlight reaching the surface, while plant communities reflect only 10 or 20 percent. The percentage of reflected sunlight is an important enough number to warrant a name of its own. "Albedo" refers to the fraction of sunlight not absorbed by a surface.

As you might expect, leaves get hot when they absorb sunlight. Leaves contain water, and heating them causes the water to evaporate. As mentioned earlier, it takes a lot of energy to evaporate water, so evaporating even a little water provides a plant with a significant amount of cooling. Plants contain water to evaporate—even the ground in green fields has water to evaporate, which helps the soil remain cool. The combination of leaves shading the ground and evaporation explains why the green meadow stays cooler than the desert.

Surfaces heated by the sun transfer their heat to the air in two ways. When the wind blows across the surface, convection causes warm air to rise. Just as blowing across soup is a quick way of cooling it, wind blowing across a surface is a highly effective cooling mechanism. The blowing mixes heat from the surface with a large volume of air. If the wind had no mixing mechanism, wind cooling would be only marginally effective. With no mixing, air very near the surface would soon become about as hot as the surface itself, and would be unable to carry away more heat.

The wind becomes turbulent as it blows; it develops a complex pattern of eddy currents that mix and diffuse the heat. This is the same mechanism that causes plumes of smoke to diffuse and mix through the atmosphere. Such diffusion, however, is a slow process. For diffusion over distances of a few feet, the heat or smoke spreads only a couple of feet per minute unless there is a high wind. Still, that means diffusion is occurring at a rate of 120 feet per hour, or about 1,200 feet during ten hours.

The 1,200 feet figure is not too far from the depth of the cool layer that develops above the earth's surface at night. Most people don't realize it, but the full depth of the atmosphere does not cool at night. Only the layer that consists of the lowest 3,000 feet or less cools down. This layer, which heats and cools each day, is called the boundary layer. When the ground cools at night, it becomes cooler than the air directly above it. The air then cools because of its contact with the ground, producing a nighttime inversion.

Here is the tie-in with global warming. If global warming causes increased wind—as it probably will—complex questions arise. More wind would increase the rate of heat and water vapor

diffusion from the surface. That could either increase or decrease the depth of the nighttime inversion and how much water the soil loses.

Convection is the other way in which the ground transfers heat to the air. Air expands when heated, becomes buoyant, and rises. This process is called convection. Air rises above a fire because gases in the fire get hotter than the nearby air, expand, and float upward, carrying the smoke along. Smoke seems to rise of its own momentum, but it is really being carried upward by hot air. The same thing applies to solar-heated ground. As the ground heats air, the buoyant air rises upward until it is no warmer than the air around it. Since it rises straight up, convection can transport heat higher and more quickly than the turbulence in the wind.

Convection is critical to transporting energy upward through the atmosphere. Water vapor in the air carried upward through convection can condense and form convective clouds. Some scientists suggest that convective clouds may provide a strong negative feedback that actually controls global warming.

TEMPERATURES AT VARYING ALTITUDE

Temperatures get colder as altitude increases. This is true at least for the troposphere, the lowest 50,000 feet or so of the atmosphere. Higher up, in the stratosphere, temperatures increase at higher altitudes. A thin layer called the tropopause separates the troposphere and the stratosphere. Temperatures remain constant at all altitudes within the tropopause. Radiation drives atmospheric temperature, but in ways you might not expect.

First consider the troposphere, where temperatures cool at higher altitudes. Solar radiation heating the ground drives these temperatures, in a process that resembles heating a pan full of water on the stove. The burner heats the water from the bottom. As the water starts to heat, you can see warmer water circulating upward from the bottom of the pan. Water temperatures are hottest at the bottom, where the burner is, and coolest toward the top. Were it not for warm water rising from the bottom, the bottom would just keep getting hotter.

The sun heats the ground just as the burner heats the bottom of the pan. Air then gets heated by the ground, expands, and rises. How high depends on how hot it is. As the air rises, the lower pressures at higher altitudes allow it to expand. When air expands, it cools. The warm air also gets mixed with cooler surrounding air as it rises. Thus the warmest air, like the warmest water in the pan, is located near the bottom. As altitudes increase there is less and less warm air, so temperatures keep getting cooler and cooler. We can say that temperatures fall with altitude, but it would be better to say that air at lower altitudes is warmed to a greater degree.

Air warmed at the surface can only rise so high. Eventually, the cooling due to its expansion and its dilution with cooler air takes its toll. The warm air from ground level is cooled to the same temperature as the surrounding air. It is no longer buoyant, so it can rise no higher. The limit to how high even the warmest air from the ground can rise is

also the lower limit of the stratosphere. Beyond this point, there is no means of warming the air from below. Temperatures stay the same as altitude increases. This is the tropopause, where temperatures remain unchanged as altitude increases.

How rapidly temperature decreases with altitude has little to do with radiative processes. The rate of temperature decrease has to do with the mixing of air, and with the cooling of air through expansion. Changing the amount of greenhouse gases in the air will not directly affect the rate at which temperature changes with altitude. This rate is generally the same in the tropics as it is in the polar regions—which means global warming should have little effect on temperature gradients at lower altitudes. Temperature gradients are important because they play a key role in cloud formation and rainfall.

In the stratosphere, ozone is the driving force behind temperatures. To put it simply, ozone absorbs the ultraviolet radiation from the sun. Heat from those ultraviolet wavelengths is what causes the temperature to rise. As Figure 4 shows, there is not all that much energy in the ultraviolet, but heating the thin air at these altitudes does not require a great deal of energy. At 100,000 feet, air density is 100 times less than it is at the surface. Without much air to heat, a little energy causes a significant temperature rise.

Ozone is not the only molecule that absorbs ultraviolet radiation in the upper atmosphere, but it is the most important one. Ozone's absorption of ultraviolet from the sun is what warms the air in the stratosphere. As the solar beam penetrates deeper and deeper into the stratosphere, less and less radiation remains at the wavelengths where ozone absorbs most strongly. Radiation decreases because the ozone keeps absorbing it. Before the beam reaches the tropopause, absorption is essentially complete; little ultraviolet remains. Ultraviolet absorption by ozone can no longer heat the air, because the ultraviolet is depleted. Somewhere along the way, the concentration of ozone starts decreasing, too. This happens because ultraviolet light is needed to make ozone. With almost no ultraviolet and less ozone, there is no longer enough activity to change air temperatures.

Within the tropopause, there is a limit to how high air warmed by the ground can be raised through convection. The limit means warming from the bottom is cut off. The lack of ultraviolet and the depletion of ozone means that warming from the top is cut off. The tropopause is essentially the no man's land that divides the atmosphere. Convection rules below; absorption reigns above.

Absorption by increased concentrations of greenhouse gases keeps the ground from cooling, and thereby causes the lower atmosphere to warm. The same does not apply at very high altitudes. Strange as it may sound, increased greenhouse gases should cause the air at very high altitudes to cool. Less infrared is coming from below to warm the upper air. Doubling carbon dioxide and methane should cause temperatures at altitudes above 150,000 feet to drop by 20°F or more. Smaller temperature decreases should extend all the way down to the tropopause.

Temperatures above 150,000 feet are more stable than surface temperatures, so scientists who are trying to learn whether greenhouse heating has already started should be looking at high altitudes. The temperature decrease should be larger than the increase they are looking for at the surface. There is another advantage of measuring at high altitudes—the temperatures there vary less from point to point or from year to year.

Of course, the story then becomes a difficult one to headline. Newscasters might find it hard to explain why the cooling up high proves that warming has set in at lower altitudes. The impact of this news is global, rather than merely visual.

▲ 4 ▼

Computer Models of the Atmosphere

If they all got the same answer, we might believe them.

Scientists are conservative people. Experience teaches them humility again and again. Too many times, they think they know the answer to a problem, only to find that the answer does not work. Such experiences teach scientists to question the obvious, to mull over answers, to wait for new results. Of course, all scientists have opinions, but most are wise enough to avoid declaring their ideas to the press. Some, however, are not. Regarding global warming and the ozone hole, a few scientists have taken adamant public positions on opposite sides— stronger positions, in fact, than the scientific evidence can support. One side says, "The end is near." The other says, "Don't worry, be happy."

Who is right? No one knows, because science has limitations and all scientists are guessing. Some think they know the answer, but too much about atmospheric processes remains unknown. Because of the unknowns, scientists cannot prove that what they think *should* happen will happen.

One other aspect of science makes the whole scenario a little scary. To make predictions for the future, we are forced to use computers. Finding how much the world will warm involves so many processes that no person—nor any group of people—can work through the problem and arrive at an answer. Only a computer can deal with such complexity. As a result, scientists who run models must instruct computers, telling them how to calculate. A major problem lies in the fact that they have no sound way of learning whether the computer calculates what they told it to calculate. Perhaps the computer program has a bug in it, causing it to calculate something other than what the modeler requested. No one can do the calculation by hand to check the computer's answer, because it is too complex.

Only a fraction of atmospheric scientists work with models. Ask the average scientist what the outcome of increased greenhouse gases might be, and you have asked a question beyond his or her ability to answer. There are so many processes and possibilities that the human mind cannot visualize and sort through them all. The average scientist can only state his opinions and repeat what he has been told that models predict. Computers, then, are providing the best answers, and we have no alternative. We are therefore in the hands of the computers. They have not taken us over in a literal sense, but they do rule parts of our lives. If the computers say we should cut back on carbon dioxide emissions, dare we defy them?

Climatic models are much like economic models. Sometimes good answers come out of economic models. But at other times they predict economic recoveries that fail to appear or forecast recessions that turn out to be less severe than predicted. Economists and economic models are often scorned, but businesses cannot afford to ignore them. Toymakers have to decide in advance how many toys to make for Christmas shoppers. The person stuck with making the decision may not believe what the economic models predict, but the model results must be taken into account. They will probably be wrong to some degree, but not as wrong as a wild guess—at least not on the average.

People tend to listen to computers because they give complex answers. Right or wrong, the answers come out loaded with details, and people believe complex answers more readily than simple ones. A wife may doubt the philandering husband who says he had to work late, but the husband who makes up a ten-minute story about the computer breakdown that occurred only five minutes before the end of the day will probably be believed. It is the same with computer output. We tend to think that the detail must have some basis in truth. After all, we and the people who put the programs together are incapable of proving the details wrong. So we have no

choice but to accept the model predictions that the computer spits out.

This view involves walking a narrow line—the line between accepting and rejecting model results; the line that says model results are probably bad but not as bad as an expert's guess. The idea here is to give the reader some appreciation for the shortcomings of models, without contending that the models are wrong. Models provide the best possible projections of what future climates will be. We must respect their results, yet understand their fallibility.

To walk the narrow line, the following approach seems best. First, we will discuss more of the shortcomings of models, to get all the problems in the open. Otherwise, cautions about model limitations would keep popping up—often enough to sound like I'm waging guerrilla warfare against them. That is not the aim of this chapter. The aim is to give both the results of models and an understanding of the factors that might make them inaccurate.

DIFFICULTIES IN MODELING

Lack of computing power is the number one difficulty in modeling. Modern computers are big and fast, but not big and fast enough. Ask an expert when computers will be up to the job of completely modeling the atmosphere, and the answer will inevitably be: "Never."

Right now general circulation models, or GCMs, break the world into blocks, perhaps 300 miles on a side. The model never attempts to calculate what happens inside this block. Everything within the block is read as being at the same temperature, with the same amount of rain falling and the same wind factors prevailing throughout the block. Models calculate the effects between blocks, but not what happens within a block.

If this 300-mile block were to be replaced by a block only a mile on each side, we would have 300 times 300, or 90,000 times, as many blocks. That means more than 90,000 times as many calculations. Models calculate at fixed time intervals. With a GCM that has 300-mile resolution, the time interval between calculations may be 30 minutes. To utilize a one-mile resolution, we would need to cut time steps to 1/300 of what they were, completing a time step every six seconds. One-mile resolution would still be too coarse to include tornados and small thunderstorms, but the increased resolution means it would take over 300 times 300 times 300, or 27 million times, as much computer power.

Why over 27 million? Because when the resolution gets finer, more equations come into play, and calculating more equations takes more time and power. With 300-mile resolution, for example, even giant thunderstorms are too small to be seen in the fixed block of space used by the model. When resolution improves, things like thunderstorms, air pollution over cities, and frost in low areas must be considered.

Even if a computer came along that could power a one-mile resolution GCM, one mile would be too large. Things like wind gusts and turbulence take place on smaller scales than a mile. The improved resolution model might give better answers, but some things would still be approximations.

Models also regard the atmosphere as a system made up of layers—one above the other, like the layers of an onion. Typically, models attribute nine layers to the atmosphere. There is no vertical variation allowed within a layer. Everything remains fixed and constant within each vertical layer, just as it remains fixed within the horizontal block measuring 300 miles on a side. This, too, is a limitation.

The blocks used in models are not square blocks. The world is a sphere, and it is impossible to wallpaper a sphere with square blocks that do not overlap. Models generally use angular blocks, like the longitude and latitude lines on a globe of the world. If you look at the angular longitude and latitude blocks on a globe, you'll notice that about five degrees is a typical size for a longitude-latitude block. Using blocks with five degrees of longitude means that 72 blocks cover 360 degrees, reaching around the world. At the equator, each of the 72 blocks would be about 350 miles long, for a total length of 25,200 miles—the distance around the world. Blocks get smaller away from the equator.

Up around the North Pole, 72 of the five-degree blocks would still circle the world. Each block would be much smaller than 350 miles in the longitudinal direction but still 350 miles in the latitudinal direction. The distance around the world at a fixed latitude gets much smaller toward the poles, but longitude distances are always the same. Right at the North Pole, any block touching the pole would cover a pie-shaped area 350 miles long.

Modelers come up with a variety of tricks to compensate for resolution limitations. The tricks work, but not perfectly. Evidence concerning the effectiveness of the compensations comes from comparing different models, as modelers often do. Comparison lets them determine whether the other fellow's tricks work better than his or her own.

The comparisons are humorous to watch. Modelers perform them in order to increase their confidence in their own answers—meanwhile insisting that their answers are better than anyone else's. Model answers never come close enough to the real weather to evaluate them by comparing them with actual weather. One model gives more accurate temperature estimates, another gives better predictions of winds. Who can say which model is more correct?

Computers run the 300-mile resolution model quickly. Only about a minute of computation is enough to forecast a full day's global weather at this resolution. But when it comes to predicting climatic trends, that speed no longer seems so impressive. Why? Because climatic changes typically occur over a long period of time. A hundred years is almost too short a timespan for significant changes to occur; yet if a computer forecasts a day's weather per minute, it would have to do so 24 hours a day for about a month to calculate 100 years of climate. Even then, most climate models take shortcuts. They often leave out noticeable details—like day and night, for example. Some do not include seasons. Few include heat transfer from ocean circulation, and none properly consider the oceans themselves.

Computing power is not the only limitation. There are factors that no one has figured out how to handle. Perhaps the best example is turbulence. Turbulence determines how effectively the wind removes heat from the ground and how well-mixed the air becomes. It also provides the drag that slows wind speeds. Unfortunately, we do not understand the basic cause-and-effect rules of turbulence. The best anyone can do is come up with some statistical formulas that indicate when turbulence develops and what its average effect is. Why it develops and what it will do once it appears we do not know.

Turbulence determines how smoke spreads from a chimney. Ask any meteorologist what information is needed to predict the exact shape of the plume emerging from a chimney at a given instant of time. He or she will tell you that such a prediction is impossible. Because the cause-and-effect relationships behind turbulence are unknown, the meteorologist is correct.

Plenty of unknowns exist besides turbulence. The formation of raindrops, the creation of weather fronts, and how much sunlight finds its way through vegetation are some important examples.

Not knowing the initial conditions distorts the carrying out of calculations. The flow of air is described by what is called a hydrodynamic equation. These equations are disagreeable, because our knowledge of initial conditions must be precise if the equations are to be useful. For example, it sounds reasonable to expect that knowing temperatures to within one degree should be close enough. You might think that such a small discrepancy would cause only small errors in the answers provided by computer models. Modelers are not so lucky, however. Mother Nature is not forgiving when it comes to input data errors. We are starting to understand the mathematical implications of such errors through a new field of mathematical research called chaos theory.

Research on chaos has nothing to do with the disorder that prevails in children's bedrooms. It has to do instead with how some equations behave when they contain slight initial errors. Studies of chaos are not limited to hydrodynamic equations describing the atmosphere, but atmospheric equations provide one area of study. How long it takes before chaotic behavior develops depends on the errors in the start-up data. Consider weather forecasting as an example.

Meteorologists, like the rest of us, realize that long-range weather forecasts are seldom accurate. Beyond five or seven days, even computer forecasts are not worth the paper needed to print them. Forecasts and weather diverge because of the chaotic nature of weather and because, even when meteorological satellites are used, the input data leading to the forecast contains too many errors. Errors of fractions of a degree in temperature cause major variations from weather predictions.

Ironically, no matter how accurate the data, it is still not good enough. Someone described the situation this way: "A butterfly flapping its wings in Argentina affects the weather in Omaha." Not immediately, but eventually.

When the meteorologist cannot predict exactly how smoke spreads out from a chimney, the lack of knowledge compounds itself. If the black smoke moves this way instead of that way, it causes the air moving in the direction of the smoke to absorb a little more sunlight. Because the smoke went this way instead of that, the added solar warming added a little more *oomph* to a convective bubble rising from the ground. That little *oomph* might spell the difference between a puffy white cloud and a thunderstorm—and the thunderstorm affects future air temperatures through evaporation from the soil. All of this occurred because the puff of smoke drifted in one direction instead of another.

Oh, yes—the puff of smoke went this way instead of that way because the butterfly flapped its wings in Argentina.

Some data is simply not available to modelers and forecasters. Here is one example: Where and how clouds form depends on condensation and freezing nuclei; but information on nuclei is neither collected nor available on a regular basis. Even more fundamental than nuclei are aerosols. Weather balloons and meteorological stations do not measure how aerosol-laden the air happens to be. People who poke fun at models refer to the effect of such bad or missing data as a "garbage in, garbage out" process.

General circulation models for climates evolved out of the GCMs done for weather forecasting. This raises the question of how anyone could use such models for climate studies. If real weather and model forecasts diverge so that the actual weather and the forecast have little to do with each other within seven days, how much faith can we have in 100-year climate projections?

As it turns out, the 100-year climate prediction is better than you might expect. The rationalization goes something like this:

Real weather and forecasts stop agreeing within a few days. Consider 14-day forecasts. They would seldom be correct. However, if we average a number of weeks of the weather that occurs, as well as several of the 14-day weather forecasts, the two averages pretty much agree. It is easier to predict averages than daily weather. The climatic models are not predicting the weather on July 4, 2092; they are predicting the average temperature in the summer of 2092. That is quite a bit simpler.

DIFFERENT MODELS GIVE DIFFERENT ANSWERS

All modelers start with the same set of equations, but from that first step onward, everything varies. Different models give different answers because models are as individualistic as the people creating them. Obviously, if the models calculated exactly what the atmosphere does, they would all get the same answer. It would be comforting to think that some models calculate temperatures too warm and that some err on the side of being too cool, but we do not know that. They may all calculate temperatures that are too cool. Still, the difference between the greatest and least numbers coming out of different models gives some feel for possible errors.

People uninvolved in math and computer modeling hold the justifiable idea that physics provides a set of equations and the computer gives the exact solution. That may be the way things *should* be, but modelers take shortcuts to make up for their lack of computer power. Many processes taking place in the atmosphere are not taken into account. For example, almost no model considers dew formation. Dew comes from water condensation and compensates for much infrared cooling. The heat released when dew forms has to go somewhere in the system, and it must return from somewhere else when the dew evaporates. Some models ignore this energy. Others may compensate by combining dew formation with some other process. Neither of these shortcuts leads to exact answers.

Models are built upon numerical constants, such as the amount of heat coming from the sun, surface emissivity, numbers defining how much turbulence the surface induces, other numbers dealing with heat transfer from the surface, and so on. The list of such numbers seems endless. Well-known numbers exist for most of these physical constants, but models may not use them. Instead, modelers do something called "tuning the model."

"Tuning" means adjusting some or all of these physical constants to alter the behavior of the model, in order to arrive at more realistic answers. Tuning compensates for processes omitted or sloppily calculated in the model. As an example, turning down the solar constant, the heat coming from the sun, might partly compensate for the effect of aerosols that scatter sunlight back into space.

Even if two models started out the same, they would no longer provide the same answers after tuning, because different users tune differently. There is no right or wrong way to tune a model. It is similar to tuning a piano by ear—every expert does it a little differently. Non-experts may do it much differently and get truly wacky

results. Getting bad answers with models is so likely that almost no modeler will loan out a copy of his or her program. If the copy were published, the bad answers would make the person who created the model look foolish. Modelers themselves get bad answers, but they have the good sense to throw out the wacky ones. Some go so far as to say that an expert is one who knows how to hide mistakes.

These are the little differences between models. *Big* differences happen when a different set of equations is used. For example, many models do not allow the the atmosphere to change the temperature of ocean water. Others let a surface layer in the ocean warm or cool, and a couple of really big models include interaction within the ocean beyond the surface layer.

Clouds provide another example. There are two or three ways in which modelers can let clouds redistribute moisture in the upper atmosphere. A couple of models even let the cloud's reflectivity change as the water content of the cloud increases. Some models are now trying to relate cloud character change to climate. Unfortunately, we have little idea of how cloud character really changes with climate.

A reasonable person might assume that putting more details in a model would ensure better answers. Strangely enough, it does not seem to work that way. Even experts have a difficult time distinguishing the output provided by a highly detailed model from the output of a very simple model.

Inserting small changes in models often causes big changes in results. In his paper, "Some Coolness Concerning Global Warming," Richard Lindzen recounts an incident involving the climate model at the British Meteorological Office. According to one run of the model, doubling carbon dioxide would cause the world to heat up by 9°F. By changing just the ice content of layer clouds, the warming decreased to less than 4°F. The small change in the model cut the predicted global warming by over one-half.

LITTLE MODELS TO BIG MODELS

Svante Arrhenius, back at the turn of the century, used nothing more complicated than a pencil to calculate how much warming would be caused by doubling the concentrations of carbon dioxide. His answer was 7 to 11°F. He assumed that the entire world would warm the same number of degrees. In effect, Arrhenius built an early climatic model.

Today, the answer generally comes out the same. The calculation proceeds in this way: Doubling carbon dioxide would cause the air to emit more infrared radiation. Each square yard of the world's surface would pick up 3.3 watts of this increased emission. That radiation would warm the ground until the ground became hot enough to emit compensatory heat. Compensation occurs when the ground is hot enough to get rid of the additional 3.3 watts per square yard it receives from above. Modern computations indicate that doubling carbon dioxide, if everything else remained constant, would warm the world by 2.2°F. That calculation also comes from utilizing a small model, albeit one that incorporates knowledge about radiation.

Now we can make the model a little larger. When the world heats, the air can hold more water vapor, and this causes positive feedback. More water vapor emits more radiation to the ground, just like increased carbon dioxide emits more radiation. That means more global warming, so emission by water vapor and carbon dioxide combined causes temperatures to increase by almost a degree, to 3.1°F.

Increased infrared emission is not the only thing water vapor causes. Water vapor also absorbs radiation from the sun. Water is different than carbon dioxide in this way, because carbon dioxide absorbs essentially no radiation from the sun. Absorbing solar radiation also causes more global warming. Taking absorption of solar radiation into account, global warming should rise by 3.4°F. That is less than half what Arrhenius calculated, but he did a great job considering what was known at the time. Surprisingly, his results come closer to what large models predict than our 3.4°F does.

When it comes to warming by water vapor, the best answer is still incomplete. Warming the air also means a change in cloud cover. A personal computer can run all the computations up to here. From this point on, big models and big computers come into play, but even the big models cannot calculate how much cloud cover will change. Doing so would take models more complete than any now existing and would eat up too much computer time. We have made progress in the last 100 years, but we still cannot get the correct answer.

Once we turn to bigger models, the idea that the whole world warms by the same amount must be eliminated. Some areas will warm more than others, and the big models indicate how much warming occurs as well as where it happens. For our purposes, it is best to stick with a single number, so we can compare results from different models. That single number is called "average heating," and we calculate it by averaging the heating all over the world. Other things come into play in the big models—things like rain, snow, and wind—so they will provide different answers than the simple models.

Table 1 shows the results from different GCM runs. The Intergovernmental Panel on Climate Change collected this set of data to illustrate the results indicated by different models if carbon dioxide keeps increasing. The models assumed that concentrations of atmospheric carbon dioxide instantly doubled. This is not very realistic, but it does greatly shorten the necessary computation.

As the table shows, these models come from research organizations around the world. The table shows the number of levels used by each model as well as whether the model included a diurnal (day and night) cycle. The table also indicates the

average increase in temperature and the percentage increase in precipitation. It breaks the models into categories that reveal something about the assumptions the models employed.

The most striking thing about the temperature changes in Table 1 is the lack of any pattern. As models become more complex, there is no clear pattern of increase or decrease in warming. Even the final, high-resolution category gives about the same

Group	No. of Layers	Diurnal Cycle	Change in Temperature (°F)	Change in Precipitation (%)
A. Fixed, zonally averaged cloud; no ocean heat transport				
GFDL	9	N	3.6	3.5
B. Variable cloud; no ocean heat transport				
OSU	2	N	5.0	8
OSU	2	N	7.9	11
NCAR	9	N	7.2	8
GFDL	9	N	7.2	9
C. Variable cloud; prescribed oceanic heat transport				
AUS	4	Y	7.2	7
GISS	7	Y	7.0	N/A
GISS	9	Y	7.7	11
GISS	9	Y	8.6	13
GFDL	9	N	7.2	8
UKMO	11	Y	9.4	15
UKMO	11	Y	4.9	6
UKMO	11	Y	5.8	8
UKMO	11	Y	3.4	3
D. High Resolution				
CCC	10	Y	6.3	4
GFDL	9	N	7.2	8
UKMO	11	Y	6.3	9

KEY		
	AUS	CSIRO, Australia
	CCC	Canadian Climate Center
	GFDL	Geophysical Fluid Dynamics Laboratory
	GISS	Goddard Institute of Space Sciences
	NCAR	National Center for Atmospheric Research
	OSU	Oregon State University
	UKMO	United Kingdom Meteorological Office

Table 1

answers as before. Within any given set of assumptions, models from different institutions give significantly different results. Categories A and B do not factor in ocean heat transport, while C and D do include ocean heat transport. But even something so important seems to make little difference. It is not that ocean heat transport is an insignificant factor. Tuning and compensation have simply offset the effects of ocean heat transport.

The ability to get around something as important as ocean heat transport suggests that modelers can make their models spew forth virtually any desired answer. Do they deliberately manipulate their results? In most cases, probably not. However, social pressures do exist in the world of science, and the fellow whose model predicts no global warming gets no respect from colleagues, so he might well be tempted to tinker with his model until it gave the same answer as everyone else's.

Perhaps the uniformity of the models' answers is great news. When the same result emerges regardless of the computation method, it may mean that the answer is quite stable. Maybe even bad calculations give good answers. Several such stable systems are known to science.

When it comes to precipitation, though, about all we can say is this: All models agree that precipitation will increase but disagree on how much and where. At least the increase is comforting. With higher temperatures, plants need more moisture because evaporation increases, so more precipitation would be a positive outcome.

The United Kingdom Meteorological Organization (UKMO) results in category C seem to be an honest set of results that give us some feeling for how touchy the models are. Temperature changes range from 3.4 to 9.8°F. Introducing a cloud water scheme dropped the temperature increases from 9.4 to 4.9°F. Changing

from cloud water to an ice formulation then increased temperatures back to 5.8°F. Keeping the cloud water scheme but allowing clouds to have variable radiative properties changed temperatures from 4.9 to 3.4°F.

Changing assumptions yields impressive changes in the calculated temperature increases. The UKMO results tell us just how sensitive models are to small changes in the way clouds are handled. Models used by other institutions would undoubtedly reflect equally large temperature variations if their cloud schemes were altered, and models may be just as sensitive to some other factors that sound insignificant. Large changes might result from factors that appear so trivial no one has ever considered them.

SOME PLACES GET HOTTER THAN OTHERS

Saying that the whole world warms by the same amount—or talking about average warming, for that matter—is simplistic. Still, people are reassured by simple constants when they are dealing with sets of numbers. Computers, on the other hand, give detailed and complex answers. The answers may be wrong, but they are wrong in agonizing detail. Such answers demand maps and graphics.

The year-round average warming from the three high-resolution models in Table 1 is quite consistent—warming ranges from 6.3 to 7.2°F. It is interesting to see how different this looks when laid out in detail on a map. The consistency is not nearly so great, but the generalizations that follow are possible.

Figures 1 to 6 show model results for temperature increases resulting from doubling the concentration of carbon dioxide. The first three maps show northern hemisphere winter, the second three maps show southern hemisphere winter. All models show that most of the tropical and much of the temperate regions will heat from 4 to 7°F. Such a good overall agreement is easy to overlook. Instead of seeing that all models give the same result for 75 percent of the earth, it is easy to look at the small areas of disagreement. The larger areas are in agreement, and these form most of the map.

Maps such as these are called contour maps. The idea is for the contour line to be at exactly the temperature written inside or along the contour line. Other points inside the contour lines can be either warmer or cooler than the temperature at the contour line, but not much. As stated above, all the models calculate warming of 4 to 7°F throughout most of the world. We can assume that regions not enclosed by contours will also experience this 4 to 7° warming.

All models generally agree that the greatest heating occurs in the winter polar regions. Arctic regions show more warming than other areas from December through February. Likewise, Antarctic regions warm more than the rest of the world in the southern hemisphere winter, which occurs from June through August. The UKMO result shows considerably less warming near the South Pole than the other two models. Even in the polar summers, the Arctic and Antarctic show a little more warming than the tropics.

Warming in the polar regions occurs in complicated patterns. Instead of showing all the detail, all the polar regions are grouped together and the typical temperature increase for the entire area is shown. All the models agree that the tropics will warm in the 4 to 7°F range. Winter or summer, warming is about the same in the tropics.

There is no need to start pulling our hair out over the prospect that the polar ice caps may melt. Authorities say we should forget about large-scale melting of polar caps. Any change in sea levels will come about because water expands as it warms. If we warm the world, the warmer water expands, causing a sea level increase of two or three feet.

Generalizations beyond these are difficult. To come up with them, we must consider details, and the models do not agree very well on details. In order to make the maps more readable, the shapes of hot and cold regions have been simplified. The temperature intervals were clear when the modelers started them in degrees centigrade, but converting to Fahrenheit caused temperatures to be stated in fractions of degrees. In the spirit of simplification, the temperatures have been rounded off to the nearest whole degree.

Let's compare the model results for the United States. The good news is that the greatest winter warming occurs in the north central and northeastern parts of the country. The UKMO shows considerably more warming in the Great Lakes region than the other two models. Winter warming is greater in the eastern United States than in the west, but the models agree poorly on how much warming we can expect. Come summertime, the models seem confused about warming in the United States. The UKMO model indicates that the greatest warming will occur in the west; the GFDL model calculates that more warming will occur in the Great Lakes region; and the CCC model declares that air conditioner sales should be strongest in Arizona and the north central region of the country.

From the three models, Europe gets little indication of of what to expect in either winter or summer. Warming is in the 3.6 to 10.8°F range, but the models place the warming in different locations. South America and Africa may have spots of warming greater than 7.2°F, but the

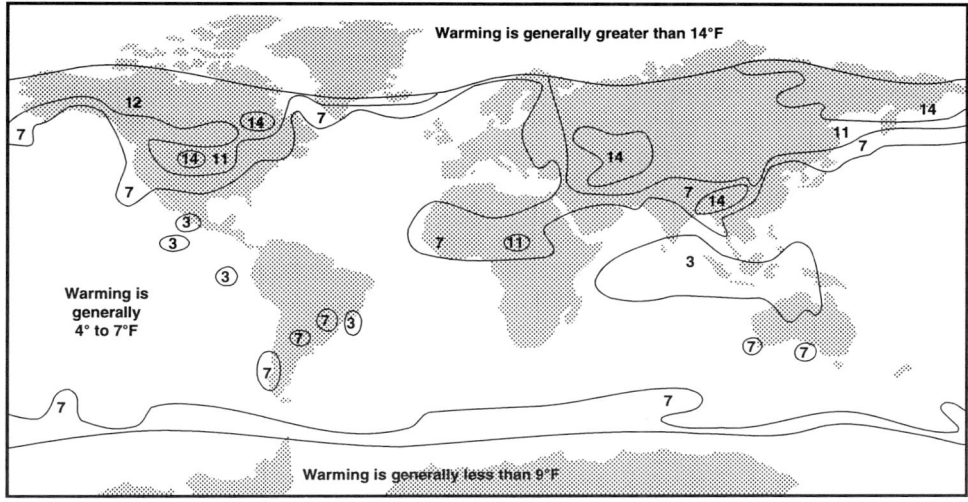

FIGURE 1   *This figure shows the northern hemisphere winter with doubled concentrations of carbon dioxide, as calculated by the Canadian Climate Centre. Temperature changes are given in degrees Fahrenheit inside or along the contours.*

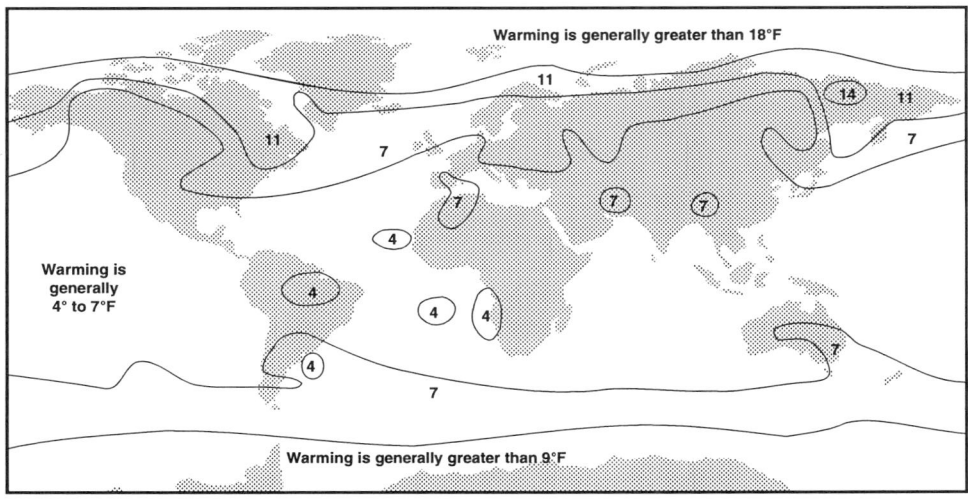

FIGURE 2   *Here, the northern hemisphere winter with doubled concentrations of carbon dioxide is shown, as calculated by the Geophysical Fluids Dynamics Laboratory. Temperature changes are given in degrees Fahrenheit inside or along the contours.*

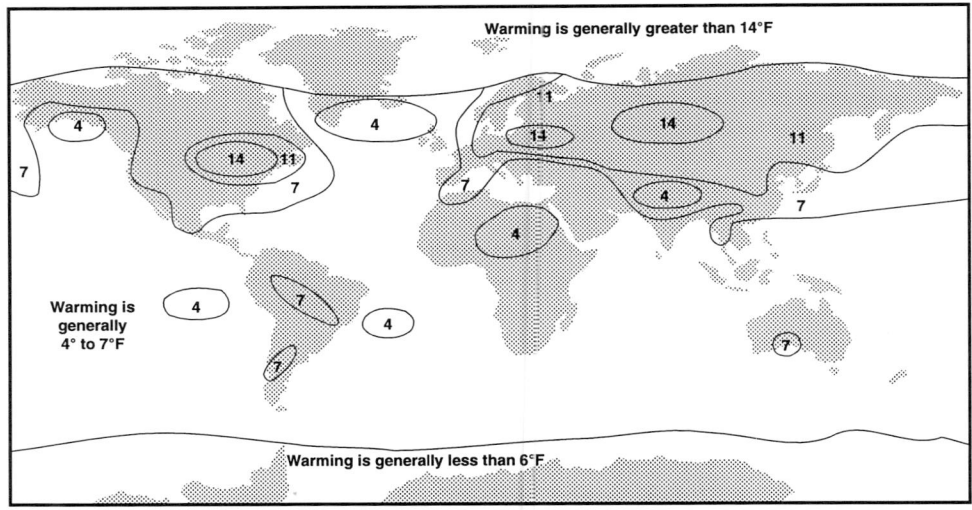

FIGURE 3 *This figure shows the northern hemisphere winter with doubled concentrations of carbon dioxide, as calculated by the United Kingdom Meteorological Office. Temperature changes are given in degrees Fahrenheit inside or along the contours.*

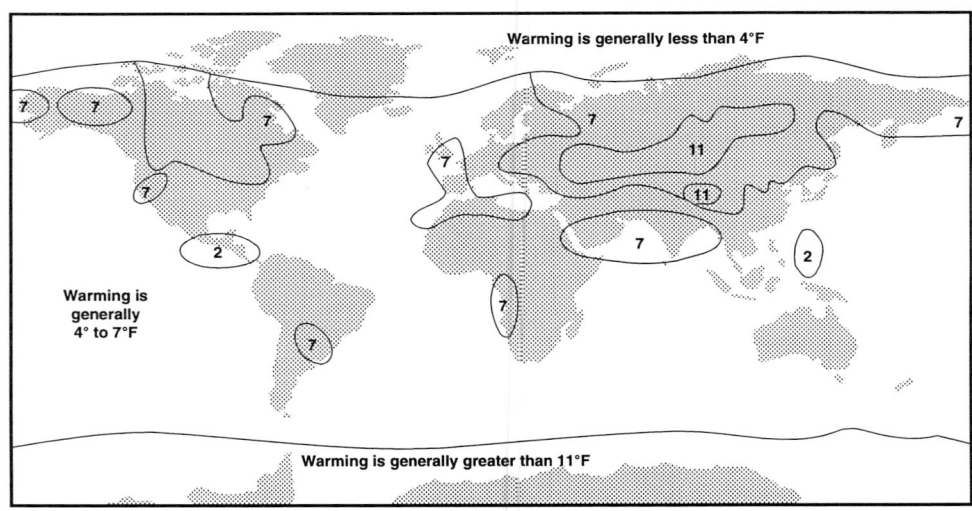

FIGURE 4 *The southern hemisphere winter with doubled concentrations of carbon dioxide is shown here, as calculated by the Canadian Climate Centre. Temperature changes are given in degrees Fahrenheit inside or along the contours.*

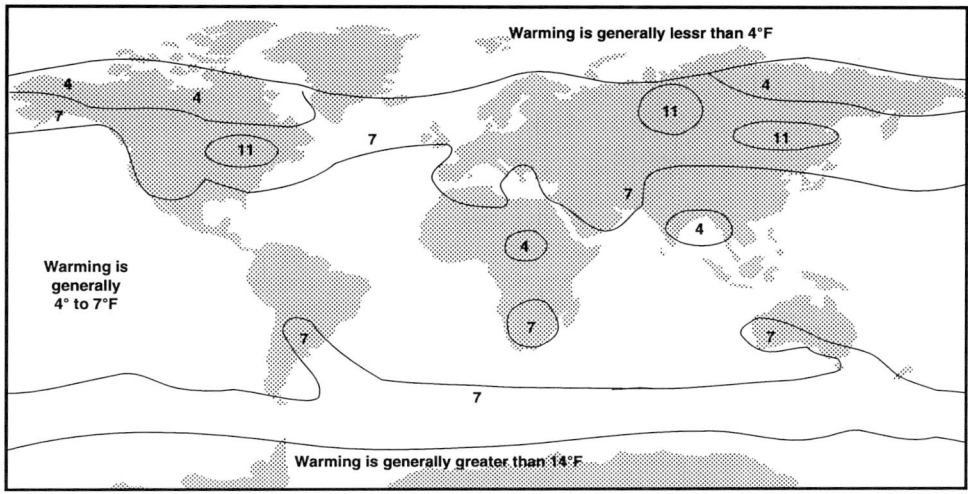

FIGURE 5 *Here is the southern hemisphere winter with doubled concentrations of carbon dioxide, as calculated by the Geophysical Fluids Dynamics Laboratory. Temperature changes are given in degrees Fahrenheit inside or along the contours.*

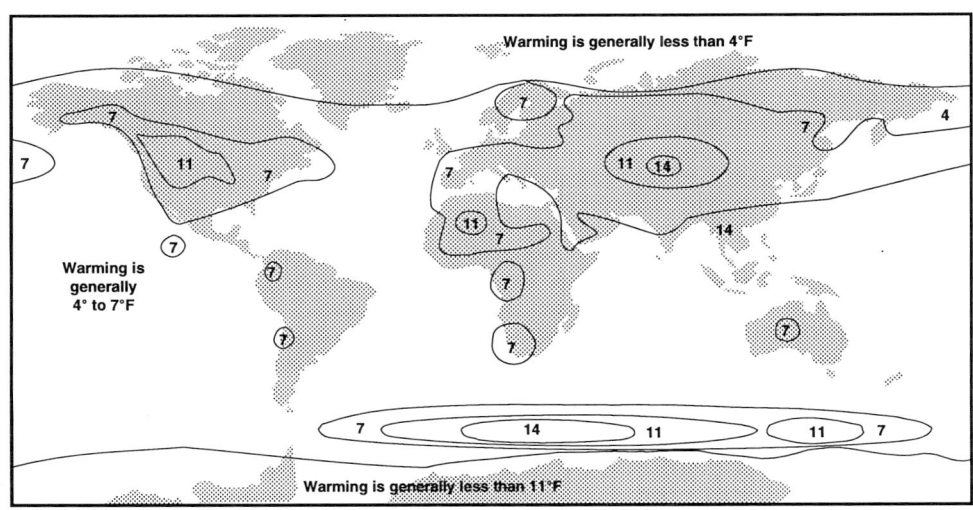

FIGURE 6 *This figure shows the southern hemisphere winter with doubled concentrations of carbon dioxide, as calculated by the United Kingdom Meteorological Office. Temperature changes are given in degrees Fahrenheit inside or along the contours.*

models show no agreement on where those spots will be.

The maps of temperature change differ from each other, but they are fairly simple to read. The maps of models that calculate *precipitation* change look like unassembled jigsaw puzzles. Almost none of the pieces seem to agree among the models. The best generalization possible is to state that no great change will occur in winter precipitation within the United States. Reasonable precipitation increases will occur in the U.S. during the summer months. The really big increases in precipitation occur in the tropical regions along the equator during both summer and winter—seasons that have little meaning near the equator. Quite a few tropical areas can expect precipitation increases of more than three inches per month. The different models fail to agree on where these areas will be located, but all models project such areas.

To carry out their calculations, the models need information on water sources for evaporation. This forces the models to calculate soil moisture, but it does not force them to do a good job. The estimates of soil moisture, which depends on land use and vegetative coverage, are a great shortcoming in the models. The models' results concerning soil moisture are not believable enough to discuss, but they generally show decreased soil moisture in the United States. That would be worrisome if it were believable.

The general agreement among the models gives us good reason to accept the notion that warming is on the way. All models agree that high-latitude regions can expect the greatest warming and that warming will be greater in winter than in summer. The models disagree on the details, but they agree strongly enough to give their projections credibility. They may calculate twice too much warming, but they might just as well calculate only half as much as they should. Representing the best estimates science can provide, these models cannot be ignored. We know that they are wrong, but no one knows to what extent. We ignore what they say at the world's peril.

THE NEED FOR BIOLOGICAL MODELS

None of the models include biological processes. There is no need to ask a model what increased ultraviolet might do, because models ignore all living organisms. The lack of biological effects is another serious shortcoming of general circulation models.

Desertification, the growth of deserts, has been a hot topic in the news. Many of the world's deserts get larger year by year. Right now, experts cannot agree on what causes deserts to grow. Deserts cannot always have experienced such increases in size, or their continued growth through time would have turned the entire world into a desert. Climate models cannot predict the future size of deserts, because deserts are more a function of vegetation than weather. Of course it seldom rains in deserts, and plants cannot grow without moisture, but the lack of rain is a feedback process. Strip the vegetation from a large region and, chances are, it will not grow back. The land will turn into desert. Dust will blow topsoil away, rains will cease, and no plants will come back to replace those that were stripped away.

Biological models are tough to build, because we do not know enough of the right kind of biology. Biologists are not very mathematical people. Some even brag about how difficult they found math and physics classes. Being untutored in math, biologists have not recognized the need for analytical results. Ask a physicist about the heat balance of a cow and he may do something that sounds ridiculous—like assuming the cow has a spherical shape. In the end, though, the physicist comes up with an equation that pretty much describes the cow's heat balance, in spite of the outlandish assumption. Ask a biologist the same question, and you probably get no more than a discourse or a table of values. The biologists want to think of life as something special, not something to be put into lifeless equations. This is fine, but with few mathematical results there is no way to build a computer or mathematical model of biological systems.

Even if analytical results were available, biological models would be difficult to create. Suppose someone wants to build a model of how a newly-seeded lawn establishes itself. Modeling the growth of grass and incorporating sunshine, shade, moisture, soil hardness, and other factors is not enough. Weeds may sprout first, sap the moisture, and shade the grass, which means the modeler must know how to model the growth of weeds, too—not just the growth of one weed, but the growth of several kinds of weeds. This is still not enough. Knowing how to model weed growth and grass growth differs from modeling the competition between weeds and grass. Life is a competitive process, and predicting the outcome in a battle for life involves much more than predicting how the grass or weeds grow when left alone.

Bugs and other animals play a part in such models, too. Moths can wipe out a giant forest. And to be persnickety, a top-notch biological model should include the human animal. Think about the problems of putting humans into a model. The model would need to predict carbon dioxide emissions by people for as long as the grass grows.

Biological models would demand more of climate models than they can deliver. Climate models describe averages—the average daily temperature, the average rainfall, the average this and the average that. Biological systems are more concerned with extremes than averages. The odd, extremely hot day may sap the strength of a plant and stunt its growth for the rest of the year. An unseasonable frost selectively wipes out several species of frost-susceptible plants. One flood can change the course of rivers, strand some plant communities high and dry, and drown other plants.

If lack of computer power limits climatic models, worldwide biological models have no chance of finding sufficient computing power. One difficulty lies in the long periods of time needed for biological systems to develop. Climatic models can assume instantly doubled carbon dioxide concentrations, then run 100 years of weather to produce climatic averages. However, it takes a mature forest much longer to develop than the 100 years now used in climate models. Weather changes, but biological systems evolve. Shortcuts that work for the weather are silly when applied to biological systems.

The weather, however, depends on biological systems. Cut the forests, replace swamps with grassland, and weather changes. Climatic models that do not encompass the relationship between weather change and vegetative cover cannot give good answers. Shortcuts like doubling carbon dioxide assume that land use and plant cover remain unchanged. This is incorrect. Those factors would inevitably change, and the answer coming out of such a model must be wrong.

Right now, no global biological models exist. This leaves us no choice but to listen to the climatic models. However, there is one great improvement that could be made in these models. Such an improvement is unlikely, because it puts a human in the modeling loop, and modelers distrust humans. But this suggestion is one way to improve the situation until better biological models come along.

We might run the climatic models, give their output to an ecologist, and let the ecologist specify what type of plants and land use to expect in the climate predicted by the model. Then the modeler could input that land use and plant cover information and continue the model run. If the climate remains the same, the answer might be reasonable. If it changes—well, that means the future climate remains an unsolved mystery.

GAIA THEORY

Gaia theory came up in the early 1970s. Some might say Gaia has no place in a book that deals with the causes and effects of environmental change, because Gaia is not hard science—or even soft science. You might call it a "squishy science."

Gaia theory claims that the collection of all living things work together to alter the environment and expand opportunities for life. It regards all life as one giant entity that attempts to make the world a more hospitable place. Scientists ignored Gaia until the late 1970s, then got tired of hearing about it and decided to expose it simply to get it off their backs. The exposure worked out differently than planned, and even accorded some legitimacy to Gaia.

A little background is helpful. To the ancient Greeks, the earth was a living goddess called Gaia. Modern Gaia's creator is James Lovelock, the British scientist who first found CFCs accumulating in the atmosphere and brought CFCs to the attention of Sherwood Roland. Without Lovelock, Roland might never have recognized what CFCs could do to the ozone. Without Lovelock, we might still be wondering what causes the ozone hole.

Scientists objected to the notion that living things cooperate to improve the environment. Cooperation implies communication, and bacteria in Guatemala send few messages to the plankton in Antarctica. The objections caused Lovelock to back off from what he states in his original book on Gaia, a book he now describes as "highly poetic." But backing off only means that he and promoters of the theory realize that Gaia needs a less poetic foundation.

Discussing Gaia helps emphasize the interconnectedness of living things and the nonliving climate, which is the theory's strong point. Most scientists focus on their own specialty and ignore everyone else's. Geologists, for example, explain the concentration of carbon dioxide in the air by noting that acid rain dissolves rocks. They seldom mention plants. Biologists, on the other hand, note that creatures adapt to their environment. They never ask how the creatures cause their environment to evolve through chemical and environmental feedback. Gaia acknowledges that all things on earth can affect, and even alter, all other things—especially climate.

One simple illustration of this involves what Lovelock calls "Daisyworld." He supposes that a few black and a few white daisies exist on a planet barely warm enough to allow them to grow. The black daisies absorb more sun, stay warmer, and thereby grow faster than the white daisies. The solar heat absorbed by the black daisies also has the side effect of warming the air a bit. That small amount of warming helps the white daisies survive, but only barely.

As time passes, the black daisies' absorption of sunshine causes the climate to become a tad warmer. Now black daisies can spread into regions outside the equator, regions that were previously too cold. After a great deal of time, the black daisies would cover the planet. Unfortunately, all the heat produced by the black daisies would eventually make the planet so hot that they could not survive. But as the climate warms, the white daisies begin to thrive. They stay cooler, because they absorb less heat from the sun. They even reflect sunlight, so the soil absorbs less sunlight. The white daisies do not drive out the black. If they did, the planet would become too cold again. Instead, the optimum mix of white and black daisies eventually inhabits the planet—white daisies in the tropics, black daisies at mid-latitudes. Living things have in effect cooperated to make the planet a more hospitable place, without ever communicating with each other.

Before Gaia, some biologists believed that complex systems like rain forests were more fragile than simpler systems. This thought was derived from models. When you put a lot of different species in a model, one or a select few of the species take over, eventually choking out most other species. Gaia indicates that such a scenario is unlikely, and that simpler systems are in fact more fragile than complex ones. The difference boils down to this: In the biologists' models, the creatures had to live with the environment as they found it. They were unable to modify it to their needs, which is not realistic.

Earth has a much different atmosphere than its lifeless neighbor planets—and a much different atmosphere than it had before life evolved. Oxygen provides a good example. The air contained a great deal of carbon dioxide when the first plants evolved, but no oxygen. Those plants had to evolve without ozone to protect them. You might say life originated in a huge ozone hole that covered the entire world. Those early plants released oxygen; oxygen formed into ozone; and sunburn became less of a problem for plants. As a group, plants had modified their environment. That same oxygen allowed animals

to evolve and serve as a complement to plants. One form of life had enlarged the possibilities of life for other life forms. Believers in Gaia point to such interactions.

Gaia claims that living things are perpetually making the planet more hospitable for other life. Humans even play a role, and our practice of farming is one example. We may have killed all the buffalo, but cows replaced them. If you count the number of cows and compare this figure to the number of buffalo that once roamed the earth, you discover that the number of cattle now supported by the prairie is far greater than the number of buffalo.

Consider the wheat fields of the Great Plains. Those wheat fields feed steers in feedlots, hens in egg factories, and many millions of people. By farming the land, people have expanded life; they have used the land to provide food for life forms beyond the prairie dogs and antelope that were displaced by the wheat fields. Whether this is an improvement or not is debatable, but one living creature—humans—made changes that increased the amount of life in the world.

The microbes that cause fallen trees to rot are not as controversial in ecology as humans are. Microbes break the tree into humus for topsoil. They don't "care" about other creatures, but they help soil form, and the soil grows plants other than the trees. Animals eat the plants and help propagate them by spreading their seeds. All of this represents living things altering their environment—not necessarily for their own good, but in a way that adds to life for other creatures.

It might seem as if, according to Gaia theory, anything we do is good because it helps other life forms. That, of course, is too simple. There is a dark side of Gaia. A group of living things can cause harm.

Black daisies out of control and spreading too quickly could make the world too hot. If they did, both they and the white daisies would die off. People can harm the environment. If we do it often and severely, we too will die off. That might be Gaia's way of solving the problem.

Once rid of us, Gaia could start over. Life could expand again because Gaia had solved an environmental problem. Given enough time, the world would evolve into an even richer system of life than it is now. Gaia tries to make the world better for life as a whole, and there is no guarantee for any particular species of life.

Gaia is an interesting philosophy. Unfortunately, it operates in hindsight instead of foresight. If we look back on life, a rationalization can be created to explain everything that happened—not unlike history, which often contends that "the right side" won every war. If we ask what might happen in the future, Gaia theory offers a simplistic answer: "I don't know, but it will surely be good for life."

Science attempts to predict the definite outcome of events. Gaia's reliance on hindsight disqualifies it as a science. It is, however, a philosophy that successfully reminds us that our destiny is contingent upon the environment around, above, and beneath us.

▲ 5 ▼

Warmer Temperatures May Increase Cloud Cover

That might save us until the oil runs out.

Clouds are the great unknown factor in global change. Many unknowns exist about the growth of clouds, but the most important one is how global change might affect cloud formation. As climate changes, will clouds increase? If so, at what altitude? If clouds increase, will rainfall also increase? At first glance, increased rainfall seems a natural product of increased cloudiness, but in fact no rule ensures that rainfall amounts will correspond to cloudiness.

Just outside Lima, Peru, lies one of the world's driest deserts. Coastal mountains block moist ocean air as it moves inland. The blocked air forms clouds that just sit there. As a result, the area is subject to nearly constant cloudiness, but no rainfall. Some might call this the world's worst weather. Still, the clouds block the sun and keep the desert from becoming as hot as it would if no clouds were present to block solar rays.

If global heating causes clouds to increase, the additional clouds may in turn either moderate or increase the heating. Everything depends on the altitude of the clouds. There is one constant, though—no matter what their altitude, clouds appear white from the top and reflect sunlight. By scattering the sunlight back into space instead of letting the ground absorb it, all clouds cool the world to some degree.

If clouds are close to the ground, their net effect is easily estimated. Near the ground, they have about the same temperature as the ground, and they emit about the same amount of radiation the ground emits. Therefore, low clouds provide a net cooling effect, because they emit the same amount of energy the ground emits but cut down on solar heating.

The effect of high clouds is more complicated. Clouds form an opaque shell that allows no infrared radiation from the ground to get through. However, the shell does not block all the sunlight; it lets a good fraction of sunlight scatter through and reach the ground. High clouds are cold, and emit less energy than the ground emits. Their net effect depends upon whether the decrease in the amount of emission, caused by their cold temperatures, exceeds their ability to scatter solar energy back into space. When decreased emission wins, clouds cause a net heating effect. Whether clouds heat or cool depends on their height and thickness.

The answer can get even more complicated than "it depends." Clouds transport energy and release it when they produce rainfall. Moving heat from one location to another can be an important factor. Rainfall is also important, because it can determine how hot the surface gets. Rain allows plant growth to cover the ground, and green fields stay cooler than bare rocks.

Scientists will never admit it, but the whole matter of clouds and their effect on climate gets too complicated for them, too. The human mind can only deal with a limited number of independent variables at one time. That is why, in business, management theory holds that one person cannot effectively supervise more than about seven workers. We can think about more things than seven by thinking of first one, then the other. However, even the most brilliant scientist cannot fully deal with the interaction of seven simultaneous variables within his or her mind.

Computers now enter the scene. They can juggle equations with any number of variables and provide answers about cloud interactions. The answer, as you might expect, is only partial, because someone has to tell the computer precisely which problem to solve. If the scientist is unable to conceive of an interaction that turns out to be critical, the computer will never be asked to attack that problem. Computers answer only the questions that people pose.

Earth Radiation Budget Satellites (ERBS) were launched as part of the Earth Radiation Budget Experiment (ERBE) in the 1980s. In reality, ERBE was not an experiment at all. Experiments involve doing something and seeing what happens as a result. ERBE did nothing; it tried only to observe what was already happening. ERBE attempted to measure incoming and outgoing radiation on a worldwide basis. Scientists desperately need a way of measuring the radiation budget in order to answer global warming questions. Satellites gather a great deal of data quickly, but scientists form committees and chew on that data for years before the results are released. The larger the program, the longer the chewing. In the case of ERBE, the first of three satellites went up in October, 1984, and the first data came out in January, 1989. Even then, the released data covered only the month of April, 1985.

Although only one month of data was provided, the numbers had great value. They provided the first confirmation of how much cooling is provided by clouds. During that month, clouds provided an average net cooling of 11 watts per square yard. This figure represents the net cooling obtained by averaging both sunny and cloudy regions. For perspective, doubling the amount of carbon dioxide in the atmosphere should cause radiation to increase by less than four watts per square yard.

Few people understand what a watt per square yard represents. It helps to consider the numbers in terms of light bulbs. According to the ERBE results, the reflection of solar radiation from clouds was 37 watts per square yard. In other words, if the world were laid out in squares measuring three feet by three feet, clouds reflected as much light as a 37-watt light bulb within each square, or a 74-watt bulb in every other square. Many rooms are illuminated by a single 75-watt light bulb. The effect of clouds would create five times the light in an average room space.

Nothing comes without a down side, and the fact that clouds block infrared cooling means that the cooling effect of clouds is not as good as 37 watts per square yard. The clouds blocked 26 watts per square yard of infrared cooling. In terms of light bulbs that is equivalent to a 26-watt light bulb on each square yard of the world, or a 78-watt bulb in every third square. The difference in the amount of solar heating blocked (37 watts per square yard) and the blockage of infrared from the ground (26 watts per square yard) gives the 11 watts per square yard of cooling. A room nine feet by 12 feet has an area of 12 square yards. Back to light bulbs: Lighting the room with two 60-watt bulbs would provide 10 watts per square yard, about the same as the net cooling of the world by clouds.

As determined by ERBE, the cooling effect was not the same everywhere. The greatest cooling occurred over mid- and high-latitude oceans. Oceans are dark and absorb almost all the incoming solar radiation; many ocean clouds are so low that they emit almost the same amount of radiation as the ocean itself. The result is little change in emission and a large decrease in absorbed solar radiation. In the tropics, the net cooling effect of clouds was essentially zero. There, the main clouds are cirrus and towering cumulus, both high cloud formations. Overall, clouds contribute more to cooling than to heating—but most cooling occurs outside the tropics.

Clouds have great potential as coolers of the earth. Whether that is good or bad depends on how cloudiness changes when temperatures warm. If warming causes less clouds, the decrease in clouds would reinforce greenhouse warming, making the weather even hotter. We must also consider the reflectivity of the ground. Deserts reflect almost as much sunlight as clouds. If rainfall decreases and deserts spread, this too might cool the world. Somehow, a proliferation of desert sounds less attractive than more clouds. Which way the climate will go is unknown. You should not be surprised that scientists are divided over how heating might affect cloudiness.

Scientists know little about clouds. Take thunder and lightning, for example. Ask a meteorologist what causes electricity to build up in a cloud, and he will probably list three or four processes. Try to pin him down concerning which of those processes really causes lightning, and a confession of ignorance eventually results. All the processes the meteorologist listed could work, but whether all processes, only some of them, or only one actually plays a part remains an unresolved problem. Ben Franklin once demonstrated that clouds contain electricity. Progress in understanding cloud electrification since then has been slow or nonexistent.

Cloud seeding is another example of our relative ignorance. For years, scientists thought they could increase rainfall by seeding clouds, and an apparently valid theory explained why cloud seeding should work. Commercial operators and government programs made all-out attempts to modify the weather. After years of debate about whether cloud seeding was working, most meteorologists now concede that seeding accomplishes nothing. The theory still looks good, but the programs did not work. Why? The best answer generally consists of shoulder-shrugging and hand-waving.

Snowflakes are another poorly understood phenomenon. Students spend their first school year cutting out beautifully symmetrical, six-sided snowflakes, and the crystallization properties of water explain why snowflakes are six-sided. How the left side of the flake knows what the right side is doing—so the left side can form in a way that mirrors the right side—has no commonly accepted explanation. The next time a child asks you why snowflakes look the way they do, you can justifiably plead ignorance. Now you can say: "Not even scientists know that."

Clouds appear white because they scatter all colors of visible light in a relatively uniform manner. If they scattered more blue light than red, their color would vary, depending on whether the viewer was seeing the light scattered by the clouds or the light passing through the cloud. The fact that clouds appear white indicates that clouds scatter all light to about the same degree. Even black clouds appear white from the top. The darkness of the cloud is a result of its thickness. It is simply too thick to allow much light through of any color.

Even very thin clouds absorb all the infrared radiation that strikes them, because liquid water and ice absorb so strongly in the infrared. All clouds scatter sunlight and absorb infrared, but this does not mean all clouds are similar; nor does it mean that all cloud cover is equivalent. Infrared radiative properties depend on the cloud's temperature. Clouds form as a result of atmospheric processes, and these processes are important in their own way. Cloud type is, therefore, more important than how the cloud scatters or absorbs radiation.

The atmosphere contains a family of clouds. There are low, middle, and high clouds. Nearly all high clouds are formed from ice particles; nearly all low clouds are formed from water droplets. Some clouds consist of a uniform horizontal layer, but others grow vertically until they reach the top of the troposphere. The number and size of droplets varies with the cloud type.

AIR POLLUTION AND CLOUDS

Air pollution greatly affects cloud formation. Clouds form when water vapor condenses into many small water droplets. It sounds simple, but a droplet has to start somewhere, and "somewhere" usually turns out to be a small aerosol particle. This is one way in which pollution plays a part. Polluted air contains more particles than clean air. Air pollution, then, modifies the weather by altering cloud formation.

Several years ago, the Illinois Water Survey looked at rainfall downwind of the Chicago and St. Louis metropolitan areas. They discovered a pattern of changing weather. Downwind of the metropolitan areas, rainfall had increased. Something in the air had caused more clouds to form and more rain to fall. Just which pollutants in the air caused the weather modification remains a mystery, but weather records prove that the change exists.

It might sound as if clouds should form when the humidity gets high enough for a collection of water molecules to bunch up and stick together. But cloud droplets seldom or never form from water alone. Unfortunately, water molecules need a surface to stick to; they do not stick very well to each other. If water has nothing to condense onto, the humidity can exceed 100 percent. If particles are present, droplets start forming before the humidity even approaches 100 percent.

The desert and mountain areas of the West usually have clean air, and boast of visibility exceeding 20 miles. If the humidity goes up, however, that wonderful visibility range goes way down. The number of particles in the air has not changed a great deal, but on humid days the particles grow larger because they take on a coating of water. Larger particles scatter more light and make the air look hazy. Water can start condensing onto aerosol particles when humidity is 30 percent or less.

Some particles attract water more easily than others. Common salt provides an example as near as the kitchen table. As you've undoubtedly noticed, the salt in salt shakers forms lumps on humid days. The lumps form because individual grains of salt attract a coating of water from the air. The wet grains then stick together, forming lumps. On unusually humid days, grains of salt lying on a table attract enough water to turn into liquid droplets.

Air containing many aerosol particles looks hazy because the aerosols scatter light just as cloud droplets do. When clouds are low, they provide a cooling effect that reduces global warming. Dirty, polluted air is usually low in the atmosphere, and it scatters light. The aerosols in air never scatter as much light as a thick cloud, but they may scatter as much as a high, thin cirrus cloud. Dust and aerosols in the air behave like clouds. They look terrible, but they help cool the world. Some have shouted "Eureka" at this point, and have told the world that "pollution is the solution." They claim that, by polluting the air, we have counteracted greenhouse heating.

For the moment, let us assume that they are correct; that no greenhouse warming has occurred because pollution scatters sunlight and counteracts the increased carbon dioxide. At the present rate of release, carbon dioxide concentrations will double every 15 years. One wonders if even the pollution apologists would be willing to suggest that we should also double our pollution every 15 years. Increasing pollution in order to counteract the effects of ever more carbon dioxide in the air does not present an attractive future.

Hot scientific arguments rage about whether recent warmer temperatures are a result of greenhouse heating. These arguments produce some strange alliances. Some scientists who accept greenhouse gases as a threat have seized the "pollution is the solution" argument, adding a new twist. "Of course greenhouse heating is already here," they claim. "But increased pollution prevents us from measuring it."

Sulfates are a particularly important set of air pollution aerosols. Coal contains sulfur, which transforms into sulfates when the coal burns. Sulfates rise from the smokestack as a gas, sulfur dioxide, which later condenses into aerosol particles. These particles play two roles: They scatter light on their own, and they affect cloud formation.

Ocean regions far from Europe and Japan are known to have whiter clouds because of the sulfate releases that occur in Europe and Japan. Sulfates seem to affect droplet size and number. Clouds with high sulfate concentrations scatter more light than ordinary clouds, because their droplets are smaller. As strange as it sounds, pollution can sometimes create whiter clouds.

Now we're in a quandary. Reducing energy use might cause increased greenhouse warming. Power plants release sulfates that scatter light and reduce solar heating. The sulfates also cause more water clouds to appear. On the other hand, power plants release carbon dioxide, which causes greenhouse heating. Both effects seem to cancel each other out. A difficulty arises if we diminish the number of power plants.

Concentrations of carbon dioxide have built up over decades. If we eliminate all use of fossil fuel and carbon dioxide, concentrations will decline over a period of tens to hundreds of years. The long delay preceding the decline of carbon dioxide levels is a major worry when we consider global warming. Even if we shut down our factories and power plants, the problem will not go away for perhaps 100 years. That is why cutting back before the problem appears makes sense.

How would shutting down power plants cause increased greenhouse heating? The increase would come about because the scattering and clouds that counteract greenhouse cooling would disappear within a few days after the closure of the power plants. The carbon dioxide might not decay for 100 years, but the aerosols would fall out quickly. Their cooling effect would be gone, but the heating effect of carbon dioxide would persevere for decades. This creates an even stronger incentive for cutting back before the problem appears. If we wait until global heating becomes a problem, cutting back will not only fail to provide an immediate solution; it will actually aggravate the problem.

CONVECTIVE CLOUDS

Convective clouds form when warm air rises. The convective cloud family includes thunderstorm clouds, hurricanes, and those nice fluffy white clouds that appear on sunny days. Nature produces a variety of clouds that form through specific processes. Almost all of the cloud formation processes are involved in the formation of convective clouds. In fact, if you can describe what takes place in convective clouds, then describing other clouds becomes easy. Put another way, if global warming affects clouds, it will almost certainly affect the type of convective clouds known as cumulus clouds. That makes them a good starting point.

First let's take a quick ride through a thunderstorm. Air heated by contact with the ground rises, then cools as it rises. When it becomes cool enough, water starts to condense into droplets. Condensing water releases heat and warms the air,

allowing it to rise even higher. When it reaches the freezing level, freezing releases still more heat. Again, the new heat warms the air, making it buoyant so that it rises. The heat released through freezing allows the cloud to continue growing until its top bumps into the tropopause, the coldest region in the lower and middle atmosphere. While this cloud growth goes on, rain or hail falls out the bottom of the cloud, and small ice crystals blow off the top. The ice crystals blow away and form cirrus clouds. The cloud also transports a lot of heat energy from the surface to high altitudes.

Now let's take a slower ride through the storm: First consider the convective rise of air. On any sunny day, some small, puffy clouds appear. These are called fair weather cumulus clouds. No one has really considered the cooling effect of these clouds. They are too small to show up in satellite images, and as a result they are neglected.

Fair weather cumulus clouds form when air near the ground gets warm, expands, and becomes more buoyant than the air around it. The buoyant air rises. How high it goes depends on how hot it is and how warm the air around the buoyant bubble is as it rises. Air pressure is less at higher altitudes, so rising air expands. As the bubble expands, its expansion cools the air. The bubble continues to rise until it is no warmer than the air around it. At this point it is no longer buoyant.

Warm air holds more water vapor than cold air, which is why cold windows steam up. Air next to the window cools and can no longer hold as much water. Its excess water condenses onto the window. Something similar happens to clouds. As the cloud rises and cools, the air can no longer hold as much water vapor. The cloud water may condense into cloud droplets if conditions are right—otherwise the air becomes supersaturated.

Before water vapor can condense, it must contain condensation nuclei, small particles onto which the water molecules stick. These particles are even smaller than the wavelength of light. Once the water condenses, it releases a good quantity of heat. It takes a lot of heat to convert liquid water to water vapor, but the heat is not lost forever. Once the water condenses, it releases the same amount of heat it absorbed when evaporating.

If enough heating takes place near the ground to cause a steady stream of bubbles to rise, more than just a little puffy cloud can develop. A major thunderstorm can result. As the storm develops, this steady flow of bubbles becomes a steady flow of rising air. The rising air reaches speeds of 60 miles per hour within the cloud. The 60 miles per hour is not a horizontal speed; it is the speed of air shooting upward one mile every minute. A full-grown thunder storm is an impressive and violent event.

Several things can prevent the storm from developing, and global warming might change these processes. We are going into all this detail to illustrate how even small changes in our climate interact with other processes. Suppose, for example, that global warming causes surface winds to increase. Wind blowing along the ground keeps the ground cooler and the air near the ground well-mixed. Air is still warmed by the ground and rises, but the wind has kept the ground cooler, so the air does not warm as much or rise as high as it would with no wind. This illustrates one way that global warming could cut down on convection, convective storms, and rainfall. Less convective storms means less clouds, both convective and cirrus. Less rainfall means less moisture at the ground for future clouds. With increased winds, weather could change in ways that either reinforce or counteract global warming. We do not even know whether more wind

creates a positive or negative feedback, so no one can say how strong its effect might be.

Pollution and global warming are unlikely to affect the supply of condensation nuclei needed to form convective clouds. However, convective storms need another set of nuclei. Without freezing nuclei, which differ from condensation nuclei, cloud droplets will not freeze until temperatures fall 10 or 20 degrees below the freezing point of water. Instead of freezing, droplets become supercooled and fail to release their heat of fusion. Freezing water releases heat just as condensing water vapor releases heat. That heat provides further warming for the convective bubble, making the cloud rise still higher.

Unlike condensation nuclei, freezing nuclei can be rare over continental areas. The whole theory behind cloud seeding to make rain involved providing additional freezing nuclei. After a droplet freezes, water vapor from the air starts sticking onto the ice, making the droplets grow rapidly. This rapid growth allows the droplets to become large enough so that they fall from the cloud as rain.

Heating is more intense in the tropics, and convective storms are more important there. These storms pump a great deal of moisture into the atmosphere. Winds carry that moisture from the tropics toward the polar region, and the moisture may fall as snow or summer rain. Not all of the snowfall comes from water carried up in tropical thunderstorms, but much of it does. Water in the atmosphere from tropical storms serves as an important greenhouse gas.

Ice crystals blowing off the top of convective clouds also form cirrus clouds. One convective storm can form cirrus clouds that cover hundreds of square miles. Light-scattering by those cirrus clouds decreases solar heating at the ground. This lack of heating may mean the surface never gets hot enough for convective storms to develop.

Clearly, convective clouds play an important role in the world's cloud cover. But the interactions we have discussed illustrate the difficulty of judging what might happen if worldwide temperatures increase.

AN END RUN AROUND GREENHOUSE HEATING

Our description of convective clouds might seem unnecessarily detailed. Why not just point out that clouds block sunshine? Because having all the details makes it possible to explain why current predictions regarding greenhouse heating may be flawed.

Richard S. Lindzen is a professor of meteorology at MIT and a nonbeliever in global warming. Lindzen claims that climate models arrive at the wrong answer because they fail to treat convective clouds properly. He raises an interesting point. No climate models include convective clouds in their calculations. They cannot include thunderstorms or even hurricanes, because no model has the required resolution. Even the best models treat the world as uniform blocks of about 100 miles on each side, and no convective storm is this large. Instead, the creators of models try to compensate for the effects of convection without calculation.

Rather than talking about models, let's consider how energy gets moved around the atmosphere. Figure 1 shows the different energy transfers in the atmosphere. The incoming shortwave radiation—light from the sun—shows on the left side of the figure. Longwave infrared is on the right. The numbers represent the percent of incoming solar energy. Thus, the longwave arrow

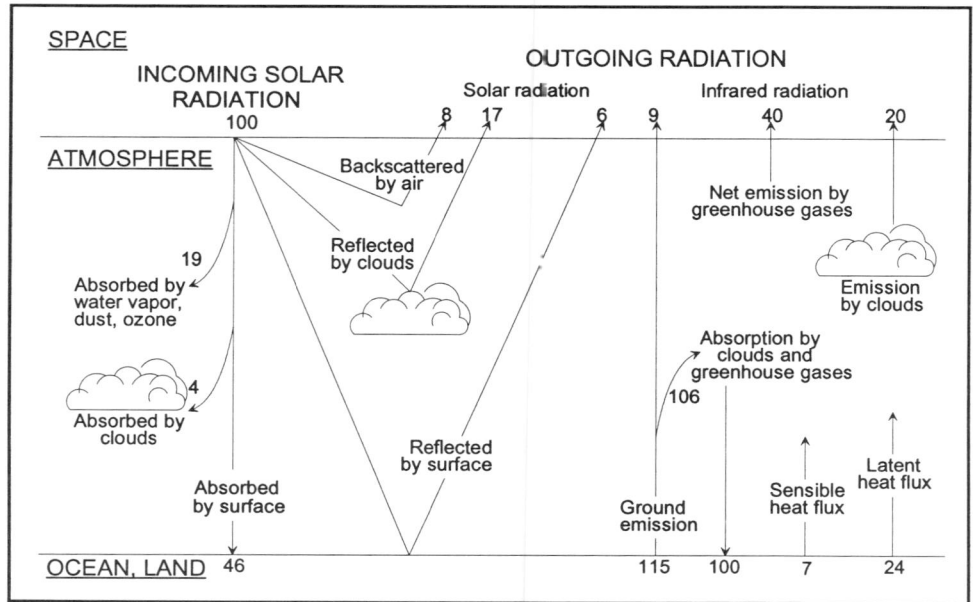

FIGURE 1 *This shows what happens to the 100 units of solar heating that reach the top of the atmosphere. The infrared emission and absorption processes are also shown. On the far bottom right are the nonradiative processes for removing heat from the ground—sensible (heat carried away by air) and latent heat (heat from water evaporating).*

leaving the ocean-land surface is 115, meaning the surface emits 115 percent as much energy as the sunshine reaching the top of the atmosphere. The thought that the ground is emitting more energy than the earth is receiving from the sun is one that deserves everyone's attention.

Figure 1 also contains two arrows that have nothing to do with radiation. These indicate the sensible and latent heat fluxes. Surfaces transfer heat to the air because the two are in contact. That transfer is called sensible heat. Radiation has nothing to do with this; it merely involves the surface warming the air as the air blows across the warm surface. Latent heat is the heat required to vaporize and freeze water. Latent heat has nothing to do with radiation either. But this is precisely what makes it so important to Lindzen's argument.

According to Lindzen and others who agree with him, latent heat provides a way to avoid much of the greenhouse blocking. Models of greenhouse heating indicate that greenhouse gases block radiation by absorbing part of the ground's infrared emissions, then reradiating some of the absorbed energy back to the ground. With more absorbing gases, more of the surface emissions are absorbed and reradiated. The ground loses less heat, stays warmer, and causes air temperatures to rise.

Convective clouds short-circuit this process. Water evaporates at the ground and rises in warm bubbles of air to form clouds. Water vapor in that air takes the heat that was required to evaporate the water along with it as it rises. When the water condenses to form the cloud, it releases its latent heat—the heat the ground supplied to evaporate the water. This is the important point. Heat was transported fairly high into the atmosphere by a process having nothing to do with radiation. Absorbing greenhouse gases

have no effect on this heat transfer.

There is more to the story. The release of heat by the condensing droplets warms the air, and this causes it to become buoyant and rise even higher. If the droplets get high enough, they freeze, releasing more latent heat at even higher altitudes. Releasing the heat at higher altitudes is an even better means of getting heat past the greenhouse gases. The higher the release, the less greenhouse gases exist overhead to block infrared radiation from escaping into space—and this heat transfer happens through convection, not radiation.

Atmospheric gases at these altitudes are not good at absorbing, which means they are not very effective as emitters, either. Clouds, on the other hand, make wonderful emitters. According to the simple-minded treatment of clouds that prevails in climate models, the cloud has to get any heat it emits from the ground via radiation. Greenhouse gases block part of that radiation, limiting the cloud's emitting capability. In the real world, latent heat supplies the cloud with energy to radiate, and this heat is unaffected by greenhouse gases.

Some numbers may help put these arguments in perspective. The total downcoming longwave radiation shown in Figure 1 is 100 percent, which represents 273 watts per square yard. Doubling the atmosphere's carbon dioxide would, according to calculations, increase the downward longwave radiation by less than four watts per square yard, or less than two percent. This means doubling carbon dioxide would increase the arrow from 100 to less than 102 (from 273 to 277 watts per square yard). That increase is a pretty small one compared to latent heat, which represents 24 percent. In essence, a tad more evaporation could compensate for all the extra carbon dioxide.

All this seems to add up to a good case,

but Lindzen offers additional ideas. He points out that most, but not all, of the greenhouse blockage results from the presence of water vapor in the lowest 5,000 or 10,000 feet of the atmosphere. A great deal of blockage also comes from water vapor above these altitudes. Convection produces rain, which tends to dry out the air at higher altitudes, removing the water vapor that blocks radiation from below. This lets the lower layers radiate more heat into space. Thus, increased convection could also counteract greenhouse heating by making all but the lowest layers of the atmosphere more transparent.

Convection dries out the upper atmosphere because air goes high into the atmosphere, where the cold temperatures convert almost all the water vapor into rain. But what goes up must come down, and what comes down is rain and dry air. Figure 2 shows how convection milks the air of its water, then sends the dried air back to lower altitudes.

Just as there is no doubt that the presence of additional greenhouse gases means less radiative losses, there is no doubt that the processes Lindzen describes take place. But that's not the argument among believers and nonbelievers in global warming. The argument concerns how much the convective processes will change as the world warms. If convection increases strongly, the world may not heat up very much, because the increased convection would counteract greenhouse heating.

CLOUDS AND CLIMATE

Convective clouds are more important in the tropics than they are at mid-latitudes. Only in the tropics and in arid areas are convective clouds the major source of rainfall. Most of the rain and snow in the United States actually comes from stratus clouds. Convective clouds grow vertically and cover only small areas. Stratus clouds are layer clouds that cover large areas.

Stratus clouds form when whole regions of air get lifted or cooled. The air, as always, must contain condensation nuclei for drops to form. Condensation, as always, releases latent heat. But things happen more slowly in stratus clouds. Lifting or cooling happens so slowly that the heat release from condensation may never warm the air rapidly enough for convection to develop. If convection does develop, the clouds become stratocumulus clouds. These are essentially stratus clouds with cumulus clouds sticking their heads above the cloud deck every so often.

Cirrus clouds are the most boring of clouds. They never produce rain or snow or do anything exciting. These are the feathery clouds that never get thick enough to block out more than a fraction of the sunshine. The fact that they are boring, however, does not mean they are unimportant.

As mentioned above, the most obvious thing cirrus clouds do is block out part of the sunshine. This blockage can limit how hot the day becomes and determine whether convective clouds develop.

The infrared effects of cirrus clouds are not obvious, but they are significant. Cirrus clouds are high and cold. They absorb emission from warmer bodies below them and re-emit it as a colder body. In other words, they perform the same function as greenhouse gases. The difference between the two lies in the blockage of sunlight. Greenhouse gases block little sunlight, but cirrus clouds block considerably more. According to most estimates, cirrus clouds contribute in a small way to global warming.

Estimates of the worldwide effects of cirrus clouds amount to wild guesses, because no good estimates exist of the thickness of cirrus cloud cover over the earth. People dealing with optics that look toward outer space often joke about the invisible cirrus clouds, meaning that cirrus clouds are present but too thin to be seen. Satellites provide the only realistic way to estimate worldwide cloud cover, and even satellites have a difficult time seeing cirrus clouds.

Some have claimed that jet aircraft create an increase in the number of cirrus clouds. Jets fly at the altitudes where cirrus clouds form. When jet engines burn fuel, they produce water vapor as well as carbon dioxide. The hydrogen in the fuel becomes water when the fuel burns, and this water is released and forms a "con trail"—short for condensation trail—behind the jet. The cold air freezes the water, which then drifts about as tiny ice crystals. Con trails often diffuse instead of evaporating. As they age, they spread out, and spreading makes them thinner and thinner as they get larger and larger. As they spread, the con trails may become cirrus clouds. If they get thin enough, they become "invisible cirrus."

This effect is most noticeable in the high, dry regions of the West. In these areas, the skies are usually dark and clear of low clouds, because little water vapor exists to increase aerosol sizes. Against this dark sky, even very thin cirrus clouds show up clearly. By mid-morning, jets from the east and west begin criss-crossing the sky. The sky that was empty of early-morning cirrus is no longer cirrus-free. As the day proceeds and the jet crossings continue, the cirrus clouds become thicker and thicker. Some theorists claim that cirrus clouds dry out the upper atmosphere. As the ice particles grow, they become large enough to settle into lower altitudes. The falling ice particles act as a pump to move water from higher altitudes to lower altitudes. Less water at higher altitudes increases radiative cooling at lower altitudes and ground level.

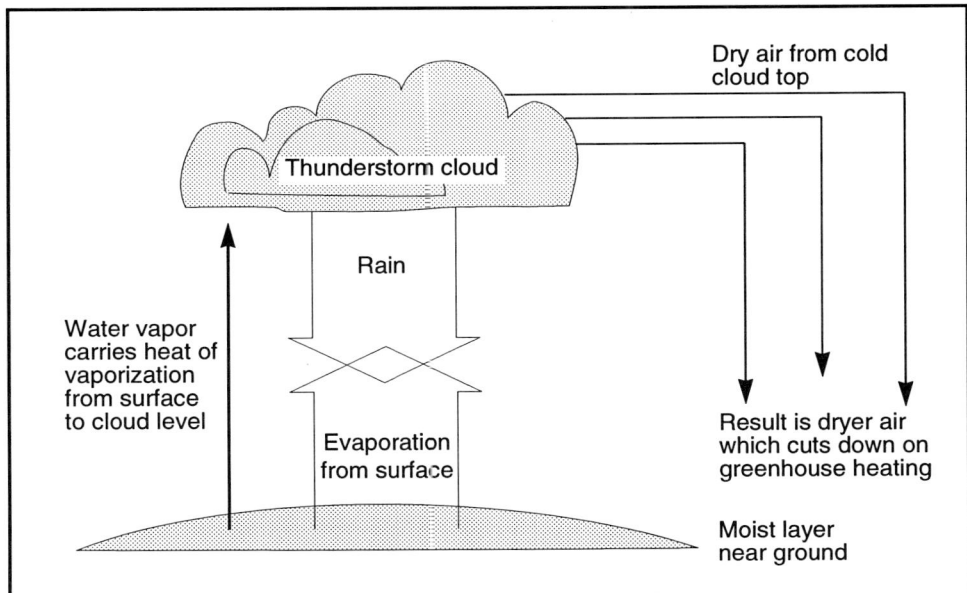

FIGURE 2 *Thunderstorms carry energy from the ground high into the atmosphere without being limited by greenhouse gases. Rising air moves sensible heat and latent heat into clouds. Once stripped of its moisture, dry air comes back down. Removing moisture reduces the most important greenhouse gas—water.*

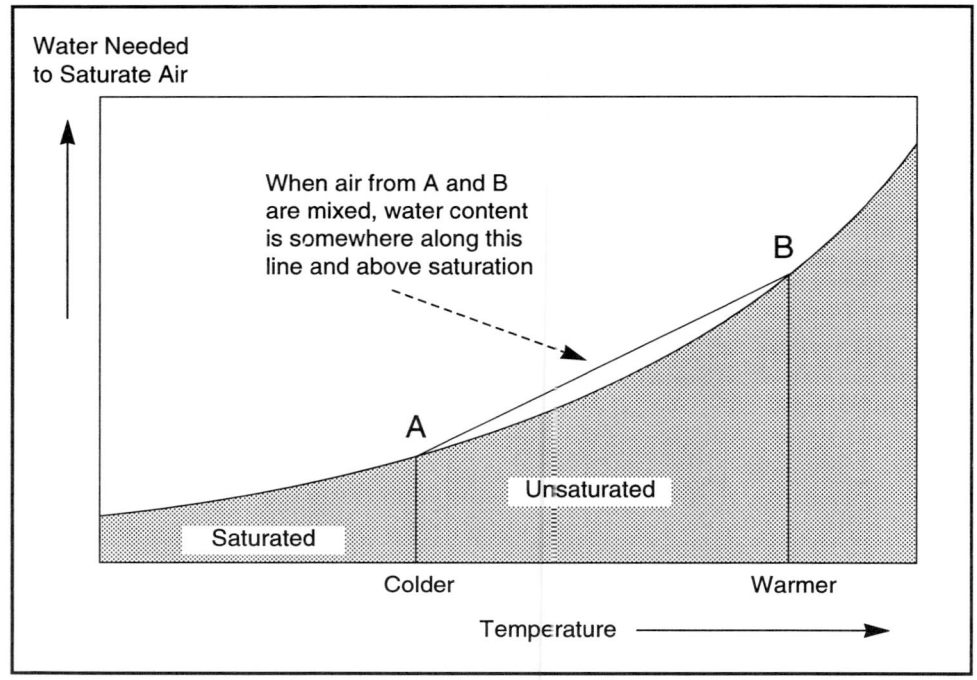

FIGURE 3   *Clouds can form when two parcels of saturated but cloud-free air combine. Mixing air from A and B produces air with water content somewhere on the straight line between A and B. This line is above the saturation limit, so the excess moisture condenses into a cloud.*

Since clouds are so important in deciding worldwide temperatures, anything that affects cloud formation is important. Clouds can form in strange ways. For example, clouds form when air cools and can no longer hold as much water vapor. Figure 3 shows the increased amount of water vapor that air can hold as it becomes warmer. The amount of water in air must always be on or below the curve. Points on the curve correspond to saturated air, air with 100 percent humidity. The important thing to notice in this example is that the curve is an upward one.

Imagine mixing a little air from point A in Figure 3 and a little from point B. Mixing air from these points results in air that has a temperature warmer than A and cooler than B. Just what that temperature is depends on how much air comes from A

and how much comes from B. In any case, the air temperature will be somewhere on the straight line connecting A and B. But all points on that straight line are above the curve that shows how much water air can contain at a given temperature.

This means that mixing two parcels of saturated, non-cloudy air causes the air to contain more water than can exist as vapor. The air is supersaturated, so the excess water will condense out of the mixed air into water droplets. In other words, it will form a cloud.

The point of this example is this: Strange things happen with cloud formation. In predicting what might happen as the world warms, we must allow for strange and unexpected cloud occurrences.

How temperatures will decrease with altitude is one of the unanswered questions about a warmer world. Meteorologists call the temperature loss with each 1,000 feet of altitude the "lapse rate." As air rises, pressure drops and the air expands. This expansion cools the air. Freely rising air would cool according to something called "adiabatic expansion." In adiabatic expansion, the temperature loss is directly proportional to altitude. That is, temperatures would decrease by the same amount each time the air rose 1,000 feet.

Temperatures do fall off at a linear rate as altitude increases, but not at the adiabatic rate. They actually fall off at about half that rate. This reduced rate of fall-off varies from time to time, but as a general rule it is approximately the same whether the location is the tropics, the polar regions, or somewhere in between. Temperatures fall off at a rate less than the adiabatic rate because the air from below does not simply rise; it also becomes mixed with local air as it rises.

Water vapor concentrations in air also fall off at higher altitudes. This reduction, too, is due to mixing. Water only evaporates from the ground, so the ground is the source of all new water moving into the air. The humid air from ground level gets mixed with dryer air from above as it rises. The mixing cuts down on humidity, just as it alters temperature.

If a new, hotter world emerges, changes could occur in the lapse rate of temperature or in the concentration of water vapor at various altitudes. These changes could happen because winds change or because the amount of convection changes. Right now, there is no reason to think that either of these processes might change. Still, the present world is the only one we know anything about.

If things warm, the earth might become different in ways that could counteract or amplify the warming. If the lapse rate or the water vapor profile changes, everyone would expect clouds to change, but no one knows how they would change. An axiom of life that also applies to the prospect of global warming is that we can expect the unexpected.

One change we can depend on is a change in rainfall. Global warming will heat some regions more than others. Models say the polar regions will heat up the most. With more warming in some regions, evaporation rates will change. Evaporation is the source of clouds, so if warming occurs, a change in clouds becomes a safe bet.

OCEAN CLOUDS

No one knows much about ocean clouds. There are no weather stations in the middle of the ocean to report daily cloudiness, rainfall, and other meteorological observations. Ships report some data, especially relating to wind and barometric pressure, but there are problems with using data from ships. They are constantly on the move, and they often report inaccurately.

Many of the weather items of interest on land have limited interest at sea. Rainfall is critical to agriculture, but hardly anyone worries about dry spells at sea. Even if they did, measuring rainfall at sea is difficult. Precipitation often comes as a mist mixed with ocean spray. The person measuring would have to decide what fraction of the water in the rain gauge is ocean water. During highly active storms, crew members rarely worry about tending the rain gauge.

Satellites can tell us something about average cloudiness, but with limitations for continental as well as oceanic clouds. Clouds often come in multiple layers, but satellites can only discern clouds. They tell us little about layers and cloud depth. The altitude of cloud tops is important to radiation, but it is information not easily obtained from existing satellites.

Low clouds frequently occur over oceans. Sometimes these clouds are thick enough to reflect appreciable sunshine. Low ocean clouds have little effect on infrared cooling, because they are nearly at ocean temperature. When they are thin, they offer the same reporting difficulty as cirrus clouds. How thick do the clouds have to be in order for us to see and report them?

A nighttime inversion develops over land, because the ground cools at night and cools the air nearest the ground. The ground cools because sunlight heats the surface during the day and infrared cooling gets rid of the heat at night.

Oceans, on the other hand, have no nighttime inversion. Sunshine is absorbed in the first several feet of water instead of just at the surface, as is the case on land. Because the heat is mixed throughout the layer, oceans have no daytime surface heating to cool off. Water surface temperatures are the same at night as they are during the day. With no hot surface developing during the day, there is less opportunity for convective bubbles to rise. Convective storms can occur at sea (hurricanes are convective storms), but their formation depends more on widespread conditions than on local surface heating.

Several other things about cloud formation are different above the sea. One is the supply of condensation nuclei. In continental air as opposed to maritime air, there is always a sufficient supply of condensation nuclei. Winds can always kick up dust, air pollution supplies more, and aerosols from plants supply others. When the humidity gets near 100 percent, water droplets start to form. At sea, a shortage of condensation nuclei can exist, and this can alter cloud formation.

For a long time, everyone assumed that salt particles should be in good supply at sea and that such salt would provide good condensation nuclei. We now know that the air under clouds at sea may be almost devoid of salt particles for nuclei. Perhaps the shortage happens because mist falls and sweeps the atmosphere clean of particles. Fortunately, there is another source of nuclei that is less susceptible to being swept away.

Above the oceans, a supply of dimethyl sulfate serves as nuclei for droplet formation. Phytoplankton are thought to be responsible for these nuclei. If so, a concern emerges. What happens if the ozone hole spreads and the solar ultraviolet wipes out the phytoplankton? Could this seemingly remote connection change the cloud cover above our oceans?

Maybe, but in the list of worries, this one can be placed near the bottom.

6

Fossil Fuels and Carbon Dioxide

Blame it on the Industrial Revolution and biodegradable trash.

Carbon dioxide is the fuel that drives the major greenhouse concerns. Several other gases cause greenhouse heating, and water is a more important greenhouse contributor, but carbon dioxide is nonetheless the central problem.

We know what causes its increase. People release 5.3 billion tons of carbon into the atmosphere each year (statistics usually report only the weight of the carbon, instead of the weight of the carbon dioxide, which would be 3.6 times as large). Methane may be increasing faster, but no one is sure just what causes that increase; humans may have nothing to do with it. There is no such mystery about the increase of carbon dioxide. It results from the burning of coal and oil.

There are secondary issues about carbon dioxide. The secondary issues deal with non-automotive sources of carbon dioxide, sinks that absorb carbon dioxide, effects on plants, and climatic trip points. These are important issues. They are the real regulators of what the world becomes if humans continue dumping carbon dioxide into the atmosphere. They give us clues regarding how much carbon dioxide can increase without causing severe consequences. Severe consequences are defined here as changes that threaten civilization, not changes that make life as we know it uncomfortable.

In order to be really sure where the earth is headed, we would need a few more planets similar to, but not exactly like, the earth. Studying how their climate differs from ours could show us what might happen if we continue burning and dumping. Computer models of climate now help forecast the future, but as we've seen, every model gives different answers, and there is no way of knowing which model is correct. Running these same models for Earth-like planets could show us which models yield accurate results and which should be trashed.

Only two planets resemble Earth—Venus and Mars—but the resemblance is not a close one. Venus has a thick atmosphere with ten times the surface barometric pressure of the earth. Its atmosphere is 98 percent carbon dioxide, and its surface temperature is several hundred degrees Fahrenheit. A look at the CO_2 concentration and the high temperatures makes putting more carbon dioxide in the earth's atmosphere sound like a bad idea. Mars has only six percent of the atmospheric pressure of the earth. Temperatures at its polar caps are almost cold enough to freeze carbon dioxide. You might say Earth has more atmosphere than cold Mars, but less than hot Venus. We might think that because Earth has more atmosphere than Mars, we experience more greenhouse heating from carbon dioxide, which helps keep the planet cozy and warm but not unbearably hot. The truth, of course, is not that simple.

The atmosphere on Mars is 98 percent carbon dioxide. Although Mars has only six percent as much air as the earth, its atmosphere contains more carbon dioxide than ours. Carbon dioxide alone, obviously, does not determine atmospheric temperature.

Carbon dioxide is the primary gas on Venus and Mars, but it is only a fraction of a percentage of ours. If the earth had no regulating system, our atmosphere would contain much more carbon dioxide, but rain and the oceans control the amount in the air. Unfortunately, these controls take 10,000 years or so to respond, and human-caused increases in carbon dioxide took place during the last 200 years. We are adding CO_2 so quickly that the controls cannot respond and restore balance to the system.

Records of past carbon dioxide concentrations come from small bubbles of air captured in glaciers. They reveal a story of ever-changing concentrations—variations that took place long before cave dwellers ignited their first campfires. Variations occur naturally; they always have, and they always will. But human activities can nonetheless cause unnatural variations. The amount of time it takes for the earth's natural regulation to respond, and the length of time between ice ages, could interact with the carbon dioxide concentra-tions. Such interactions may cause systems to oscillate—to swing eternally back and forth between extremes.

Strange as it seems, scientists cannot explain where almost half of our carbon dioxide releases go. With so much current interest in greenhouse heating, you might think that priority would be given to finding the "missing" half of the carbon dioxide, but this has not been the case. Tracing it is important for two reasons. First, whatever is taking this carbon dioxide out of the atmosphere may get saturated and could stop absorbing it, which would mean accelerated future increases. Second, if something inconspicuously absorbs half the increase, perhaps there are ways to make the substance absorb the other half, too.

Although scientists cannot account for the whereabouts of much carbon dioxide, you never find the missing gas listed in a chart. The charts always balance. Instead of admitting that a certain percentage is unaccounted for, charts place a little here, a little there, not entirely unlike an embezzler doctoring the books to cover a theft.

Part of the trouble has to do with the structure of science. No scientist wants to admit that those in his field have been overlooking something so important. Instead, some scientists insist that the missing carbon dioxide is the responsibility of someone in another specialty.

One simple and reasonable theory has been suggested to explain the missing material: With increased carbon dioxide, plants grow more rapidly, converting much of the excess carbon dioxide into plant matter. This is consistent with several pieces of evidence, but scientists are not likely to accept such an obvious solution without years of debate and study.

THE MOVEMENTS OF CARBON

Carbon is the stuff of life. Every living

thing contains carbon, but "every living thing" is too many things to talk about. Instead, it is easier to put carbon into categories and discuss the categories as a whole.

Carbon dioxide provides the motivation for discussing carbon, but carbon dioxide comes and goes, while the carbon remains. For example, carbon dioxide dissolves in ocean water. Upon dissolving, carbon dioxide molecules break up and regroup into other arrangements of atoms. Carbon goes one way, the oxygen in the carbon dioxide another. Water itself is partly oxygen, so keeping track of the oxygen from carbon dioxide is essentially hopeless.

Carbon is another matter. Several chemical species of carbon may exist in ocean water, but they are only of interest because they may eventually become carbon dioxide again. If we keep track of carbon, we keep track of the only material that could become carbon dioxide, which is the substance of real interest.

What should the carbon categories be? Some are obvious, such as the categories of carbon in the atmosphere and carbon in

ocean surface layers. Less obvious categories appear in Figure 1. Land biota are things living on land. Crumbled organic matter in the soil forms detritus. Carbon keeps getting into living things, then going back to inert chemical form and waiting to reenter a living system. We live and get counted as land biota now, but we cannot remain so forever. As organic matter decays, most of it becomes carbon dioxide that gets back into the atmosphere. Some converts into soil detritus.

Figure 1 looks a little busy, but a short road map should make it clear. Within each box is a number. That number tells how many billions of tons of carbon exist in that category. In three of the boxes, there is also an underlined number, which indicates how many billions of tons of annual increase are caused by human activities.

A billion tons is an awe-inspiring number, but including many billions of tons on the same chart can help us develop perspective. To say the marine biosphere contains seven billion tons of carbon inside living creatures should impress

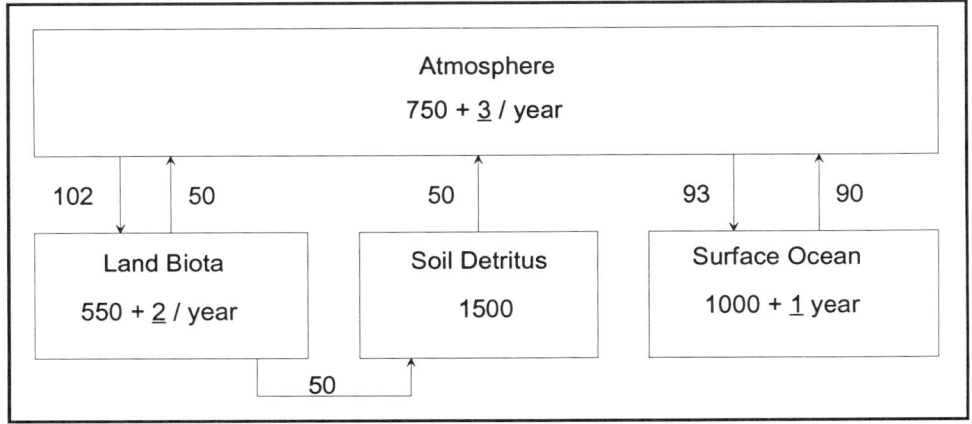

FIGURE 1 *The billions of tons of carbon in different reservoirs are shown in the boxes. The increase per year is shown as an underlined number. The numbers beside the arrows show the annual exchanges. For example, the ocean accepts 93 billion tons, but releases 90 billion tons. Inputs less outputs will agree better with the increases shown after adding the additional exchanges in Figure 2.*

everyone. Looking at the chart offsets that impression a bit, because there are 550 billion tons of carbon inside land-based living entities. Figure 1 shows that the atmosphere contains about the same amount of carbon as all land biota. Plants account for the bulk of that land biota matter, not mammals. Most of the material in soil detritus comes from things like tree roots that linger for years after the tree goes into the fireplace. The soils of rain forests harbor a great deal of this formerly living plant matter.

Besides boxes, Figure 1 contains arrows with numbers beside them, which indicate how much carbon moves each year, where it comes from, and where it goes. Sometimes the carbon moves along a two-way street. For example, each year, growing things on land take 102 billion tons of carbon from the atmosphere. Plants use the carbon dioxide to grow. Animals remove little carbon from the atmosphere, but their respiration puts a lot of carbon dioxide into the atmosphere. Animals get their warmth and energy from food, and that means they convert the organic carbon in their food into carbon dioxide.

Most people never think about what the carbon dioxide they exhale represents, even though teachers drill into us the concept that we breathe in oxygen and breathe out carbon dioxide. That is where the carbon in Figure 1 gets into the atmosphere—from our breathing. Part of that carbon dioxide comes from burning food to keep our bodies warm or to perform work. There is one other place it could come from—fat.

Have you ever wondered where the weight goes when you go on a diet? Instead of burning food, diets force the body to burn fat. When burned, the fat becomes nothing more than water and carbon dioxide. Our breath contains water vapor along with carbon dioxide, and that

is precisely where the lost weight goes—we breathe it out. Someone could develop a weak justification for shunning diets by noting that dieting increases the atmosphere's greenhouse gases. However, the combined weight of everyone living on the world totals well under a half-billion tons, so mass dieting is no threat to our carbon dioxide levels.

Three arrows in the figure are relevant to global warming. They are the arrows representing the quantity of carbon that reaches the atmosphere from land biota, soil, and the ocean. Oceans and continents put about the same amount of carbon dioxide into the air. Land-based biota take out 102 billion tons, about twice what they put back into the atmosphere. Eventually, the extra carbon from land-based creatures moves into soil detritus and back into the atmosphere. Notice that the land and its biota take in two billion tons more carbon than they release. That two billion tons accounts for where part of our carbon dioxide releases go.

Figure 2 is simply an expansion of Figure 1 that includes a few more categories. The added categories are: earth-bound carbon, ocean biota, and the deeper ocean layers. The most impressive number is the amount of carbon in the deeper ocean layers—38 *trillion* tons, or 50 times as much as there is in the atmosphere. Figure 2 also contains arrows representing the amount of carbon that people put into the atmosphere. People take the carbon from the earth-bound carbon category, which includes fossil fuels and rocks.

Large things such as trees and cows contain much of the carbon on land, but just the opposite is true in the oceans. Most of the ocean's living carbon is contained in small things—primarily tiny plankton. It takes 100 pounds of phytoplankton (a minute plant) to grow ten pounds of zooplankton (a minute animal that lives off of phytoplankton). That ten pounds of zooplankton could raise no more than a pound of fish, and most of that pound would be small-sized species. Those little fish could produce only a small fraction of a pound of fish the size we eat.

Carbon from the deep ocean forms carbonate rocks. After they emerge from the ocean, these rocks dissolve and release carbon into the atmosphere. The amounts are too small for the figure, but the formation and dissolution of rocks is the long-term determinant of how much carbon is contained in our oceans. Eventually, dissolving rocks play a part in deciding the atmospheric levels of carbon dioxide.

Rocks dissolve in the following manner: First, a little of the carbon dioxide in the air gets dissolved in rainwater. Since carbon dioxide is soluble in water and the raindrops are surrounded by air containing carbon dioxide, the raindrops absorb some of it. Once in the raindrop, carbon dioxide is no longer carbon dioxide. Together with the water, it forms carbonic acid—not a very strong acid, but acid nevertheless. And so the world experienced acid rain long before smokestacks were constructed. Of course, the acid rain from smokestacks is mainly sulfuric and nitric acids, which can be more potent than the carbonic acid in rain.

The acid in rain dissolves some of the earth's rocks after it falls. We generally refer to this process as "weathering." Weathering describes the acid's action better than dissolving, because some parts of rocks dissolve more slowly than others.

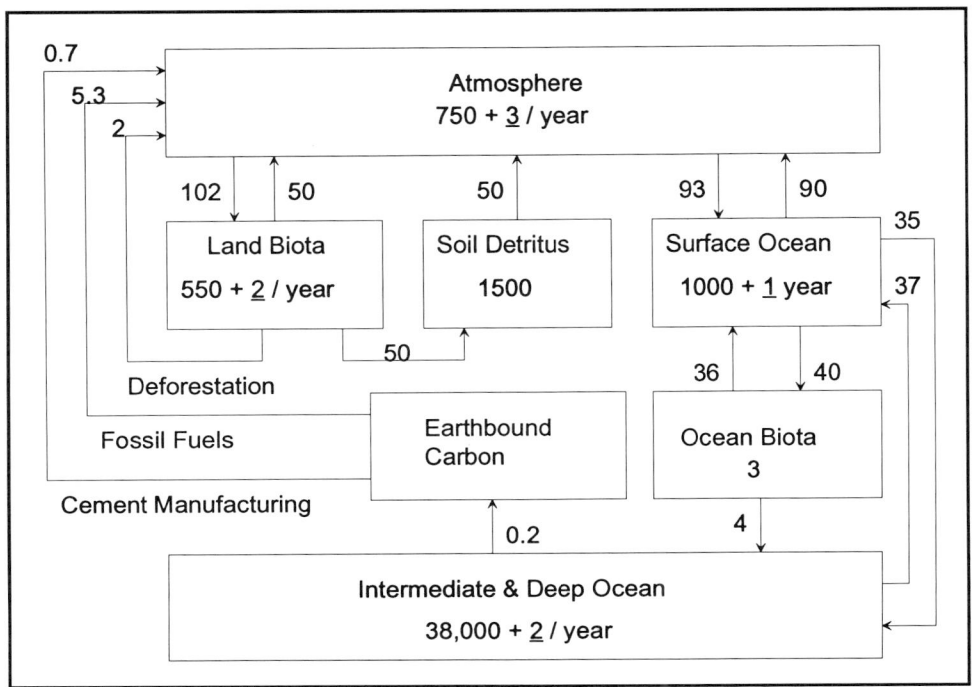

FIGURE 2   *This is a more complete carbon budget. Additional processes have been added to those in Figure 1. Exchanges and increases are still only in approximate agreement, primarily because numbers have been rounded off. Most of the numbers are only expert guesses, and different experts advocate widely different numbers.*

Life in the ocean depends on weathering. Rain runoff takes a river ride to the ocean and carries with it minerals dissolved from rocks. This runoff supplies the ocean with minerals for growth. Without calcium or silicon to build shells many life forms, like diatoms, would disappear. Other minerals are important, too, but calcium is of particular interest with regard to carbon.

If creatures did not form shells, and if those shells did not eventually settle to the bottom of the ocean, there would be no limestone formation. Limestone is mainly calcium carbonate. If the ocean is to keep absorbing carbon dioxide from the atmosphere, it has to eliminate that carbon somehow. The formation of limestone and other carbonate rocks utilizes 0.22 billion tons of carbon each year. However, carbonates are not the only way oceans get rid of excess carbon. Some of the carbon becomes shales or noncarbonate rocks; perhaps a little of it forms oil.

Now we arrive at the effects of humans. People put two billion tons of carbon into the air from land use, mainly by clearing forests. People burn enough fossil fuel to release 5.3 billion tons of carbon into the atmosphere each year—about a ton for each person on earth. This carbon comes from the consumption of oil and coal. The ton-per-person figure neglects other important carbon-releasing activities, like wood burning and agricultural burning.

The burning of fossil fuels gets most of the media attention, but this is not the only geologic reservoir we are tapping. By roasting rocks to make cement, we take 0.7 billion tons of carbon from the earth and vent it into the atmosphere. That is one-eighth as much carbon dioxide as the total released by consumption of fossil fuel, and one-third as much as is produced by clearing rain forests. However, no one accuses cement-makers of contributing to global warming.

Before humans entered the scene, carbon dioxide releases from volcanoes and the formation of metamorphic rocks might have been important, but since their combined releases total only about 0.1 billion tons, they are not shown on the charts. Why not ignore them, when the human releases are many times that large?

WHERE IT COMES FROM, WHERE IT GOES

Carbon dioxide in the atmosphere traces its ancestry back to volcanoes. An eruption may have happened in the far geologic past, and the carbon might have been recycled several times, but volcanoes are nonetheless to blame. Volcanoes spew out carbon dioxide, as well as water vapor and other, less interesting gases. Until volcanoes had released enough water vapor for clouds to form and rain to fall, there was no way for the carbon dioxide to get out of the air. At one time, concentrations may have been ten times higher than they are today.

It might sound as if the world really had a greenhouse problem in earlier times, but it did not. The high concentrations of carbon dioxide occurred before the atmosphere had much water. Remember, water is the most important greenhouse gas. The combination of reduced water and lots of carbon dioxide might have made the greenhouse warming effect less pronounced than than it is now.

After volcanoes had emitted enough water vapor for rain to fall, the carbon dioxide cycle started. The cycle essentially continued on its own until the time of the Industrial Revolution, when people began burning great quantities of fossil fuels. On its own, the world balances its carbon dioxide budget, but it has not yet been able to get human emissions under control. Our emissions began mushrooming less than 200 years ago; nature's devices require perhaps 200,000 years to stabilize things. Nature's control starts with rain.

Carbon dioxide dissolves in water, so rainwater removes some carbon dioxide from the air. The rain that falls on land dissolves carbonates from rocks, because the dissolved carbon dioxide in the water makes it acid rain. Carbonate rocks contain calcium, carbon, and oxygen. Once dissolved by the rain, the carbonates become bicarbonate ions, $CaCO_{3^-}$, which are carried to the oceans by rainwater runoff.

In the oceans, organisms that need a little calcium for their skeletons or shells capture those bicarbonate ions. The organisms convert the bicarbonates back to carbonates. Sea shells are one example of carbonates, but even microscopic creatures like diatoms live in a carbonate shell.

When the sea creature dies, its shell settles toward the bottom of the ocean. The top layers of the ocean are saturated in carbonates, so the shells do not dissolve. If the ocean bottom is not too deep, the shells settle on the bottom and eventually become carbonate rocks like limestone and calcite. Deeper down, pressure and temperature are such that the ocean is no longer saturated with carbonates. If the shells fall into the deep ocean, they dissolve before reaching the bottom. After dissolving, they become bicarbonate ions again, ready to be used in another shell when the deep water rises to the surface 1,000 years or so later.

Carbonates are not the only carbon ions in the ocean. Some carbon dioxide dissolved in rainwater falls in the ocean. This rain is also acid, and it makes the ocean itself slightly acidic. If too much carbon dioxide found its way into the ocean, the shells of dead creatures would start dissolving before they settled into the deep ocean. This would create a problem, because the ocean would become less able to absorb the carbon dioxide we produce.

Carbonate rocks formed from materials settling in the ocean may find themselves subject to heat and high pressures. If so, metamorphosis sets in and transforms these rocks into silicates. During the formation of silicate rocks, carbon dioxide is released and takes part in another cycle.

Carbon may take other detours. Some organic material from creatures living in the top layers of the ocean fails to dissolve as it settles. It becomes buried while still in organic form, and this material eventually becomes oil and coal deposits. No one seems certain whether such deposits are still forming, but modern formation seems likely. However, it takes a few million years for the formation and migration to produce oil and gas fields.

The cycle of dissolving and forming rocks provides long-term control of carbon dioxide in the ocean and air. More carbon dioxide in the air means more acid in rain, more rocks dissolve, and more calcium as well as carbon enters the ocean. The added calcium and carbon form carbonates that end up in rocks.

A short-term control cycle also exists. It could solve our carbon dioxide problem in just 20,000 years or so. The short-term cycle also involves precipitation of carbonates, but because it takes longer to work than all of recorded history, it cannot solve humanity's immediate environmental problem. The short-term system is a highly complex one that we will not examine here.

The universal question seems to be: Where does the carbon dioxide in the air come from? In order of decreasing importance, the sources of CO_2 are:

1. Surface exchange with sea water, governed by wind speed and temperature;

2. Decay of plants and exhalation by animals;

3. Burning of fossil fuels;

4. Direct release of carbon dioxide from soil;

5. Gas from volcanoes and hot springs.

What happens to the carbon dioxide when it leaves the air represents the other side of the coin. We need only a short list:

1. Surface exchange with oceans;
2. Photosynthesis;
3. Weathering of rocks and development of new fossil beds.

The first item appears to be the same in both lists. However, as shown in Figure 1, the atmosphere gives a lot more carbon dioxide to the ocean than it receives.

We tend to think about where carbon dioxide goes, but few of us wonder where it does *not* go. About half the carbon dioxide that has been released from fossil fuels since the onset of the Industrial Revolution is still in the atmosphere. The oceans are running more than a little behind in the exchange.

Things get worse. No matter what quantity of CO_2 we release, over the same amount of time nature disposes of only half of the new release. The process resembles radioactive decay, in which half the remaining element requires one half-life to decay, no matter how much happens to remain. Essentially, this means the process will take forever, whether we're talking about radioactive decay or disposal of excess carbon dioxide emissions. No matter how many times a quantity is cut in half, there is always some amount left.

Therefore, even if we greatly reduce emissions, the remaining amount will cause carbon dioxide concentrations to increase. It is not a question of how much CO_2 the world can absorb, because absorption happens too slowly to be practical. Instead, the question becomes: How much are we willing to let carbon dioxide levels rise? The people and governments of the world must answer this question, then determine how we can cut back to the necessary emission levels.

Here are some statistics that many people find mind-boggling. The 4.5 percent annual increase in fossil fuel use works like compound interest. That 4.5 percent increase means that consumption is ever-increasing, and will double every 15 years. This in turn means that, in the last 15 years, our carbon dioxide releases from fossil fuels equaled the total of all previous releases. In the next 15 years, the releases will be twice that amount.

What is the worst-case scenario? If we burned all of the world's fossil fuels tomorrow—in one day—atmospheric carbon dioxide would become ten times what it is today. Life might not survive if carbon dioxide levels increased that much, but it is oddly comforting to know that there is a limit to the potential increase.

CHALLENGING THE CONVENTIONAL

The preceding explanations are the conventional view of carbon dioxide sources, sinks, and movements. They are estimates, not measurements. In 1990, a trio of researchers decided to check out these "budget figures." The researchers put what was known about carbon dioxide concentrations into an atmospheric general circulation model (GCM) to determine how well the conventional explanations fit with the facts.

The fit left a lot to be desired. According to their results, for example, the oceans are not taking up the excess carbon dioxide. They conclude that we do not in fact know what happens to about half the carbon dioxide. The researchers doing this work were Pieter Tans from the University of Colorado, Inez Fung from NASA, and Taro Takahashi from Columbia University.

First, there is the question of what percentage of our carbon dioxide releases gets absorbed and what percentage stays in the air. From 1981 to 1987, the measured rise in atmospheric carbon dioxide was 57 percent of the fossil fuel input. In other words, roughly half the carbon dioxide we put in the atmosphere stayed there; nothing absorbed it.

Some news stories try to invoke fear by suggesting that the oceans may soon be full and may stop absorbing our excess emissions. As we see, the oceans may not be absorbing the carbon dioxide, but something else is. If the oceans—or this something else—stopped absorbing carbon dioxide, carbon dioxide levels would simply start increasing at twice their current rate, since only half the carbon dioxide is currently absorbed. That would be unfortunate, of course, but it would not represent the runaway increases most people imagine when they hear such news stories.

Here is one reason the oceans are not about to stop taking up carbon dioxide. According to Tans, Fung, and Takahashi: "If half of the cumulative fossil fuel carbon dioxide emitted since 1850 were distributed uniformly in the upper 3,000 feet of the oceans, the total carbon dissolved would increase by only one percent."

The statement reveals much about what we do not know concerning the ocean's absorption of carbon dioxide. Such small increases are unmeasurable. Given such tiny increases, it is impossible to determine whether the oceans actually absorb any of our carbon dioxide releases. We can measure carbon dioxide increases in the air, but not the increases in the ocean—not even in the top layers of the ocean.

The people who devise carbon dioxide budgets for the world have always assumed that the oceans accept the carbon dioxide. Based on back-of-the-envelope calculations, they came up with numbers they liked. As it turns out, their numbers were wrong.

Tans, Fung, and Takahashi performed their calculations by using known carbon dioxide releases and concentrations at various north-south locations around the world. Most fossil fuels are burned in the northern hemisphere, but most oceans are in the southern hemisphere. By using the computer model of atmospheric motions and absorption in the ocean, they calculated the differences in concentration levels between hemispheres. The differences upset the conventional wisdom that oceans absorb much of the carbon dioxide.

The only way the researchers could explain the difference in carbon dioxide between the northern and southern hemispheres was to conclude that oceans are not the major sink for carbon dioxide. They suggested that, instead, terrestrial ecosystems on continents absorb most of the carbon dioxide removed from the atmosphere.

Here are some numbers that support their conclusions: From 1981 to 1987, fossil fuel burning released 5.3 billion tons of carbon per year; atmospheric carbon dioxide increased by three billion tons per year. The researchers concluded that, at most, oceans could only absorb one billion tons per year. Continental ecosystems, they decided, must absorb 2 to 3.4 billion tons per year. This means that land is the big absorber of carbon dioxide.

These are startling conclusions. The researchers published their results in *Science,* a highly prestigious scientific journal, which reviews its submissions so carefully that only a third ever get published. The researchers work for well-respected organizations—NASA, Columbia University, and the University of Colorado. They did nothing radical to get their results. All of this means that the results must be respected. It does not mean that they are correct.

All we can say now is that the ball is in the court of those who say the oceans absorb a great deal of carbon dioxide.

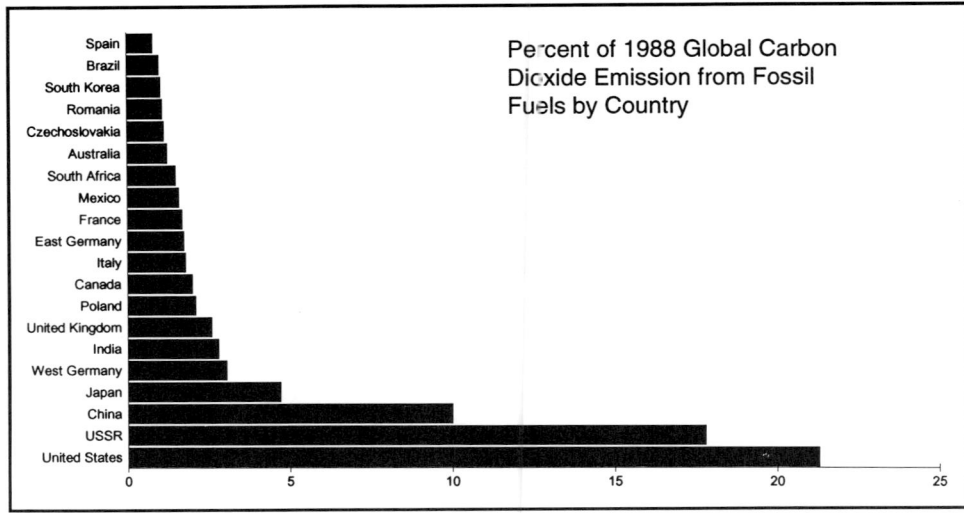

FIGURE 3 *When carbon dioxide releases are stated as releases per person, East Germany emits more than the United States. However, this does not take the amount of industrialization into account.*

Science works that way. The three researchers have pointed out the problems with existing estimates, and have arrived at estimates of their own. Now the other side will undoubtedly attempt to find flaws in the newest results. This back and forth process is one reason why scientific problems often take so long to resolve. Scientists come to many incorrect conclusions, but it takes time for other scientists to pinpoint the mistakes. Even after the errors are found, someone else has to work out the correct answer.

For the rest of us, the carbone dioxide results boil down to this: No one knows where half of the carbon dioxide goes. Scientists thought it was being absorbed by the oceans, but there is a problem with this theory. We do not know whether the excess ends up in the ocean or on the land. If it ends up on land, we have no idea where. It seems reasonable to suggest that it is utilized in plant growth, but even this hypothesis remains unproven.

As is often the case, the more we know, the less we understand.

THE FINGER OF GUILT

Peter's Quotations by Laurence J. Peter, author of the *Peter Principle*, credits Benjamin Disraeli with saying, "There are three kinds of lies: lies, damned lies, and statistics." In this section we'll look at who should be blamed for flooding the world with carbon dioxide. The only way we can decide is by consulting the statistics. Yet, as the quotation implies, statistics can be twisted.

Who emits the most carbon dioxide? The obvious answer, the one published all the time, is the United States. Figure 3 shows the statistics from 1988. The Soviet Union was the only country that approached the U.S. Since the collapse of the Soviet Union, there is no longer any country that approaches the U.S. emission rate. The next closest country, China, emitted just half as much as the United States, despite its huge population.

Let's change the question: On a per-person basis, who emits the most carbon dioxide? When we put it this way, as shown in Figure 4, East Germany was the

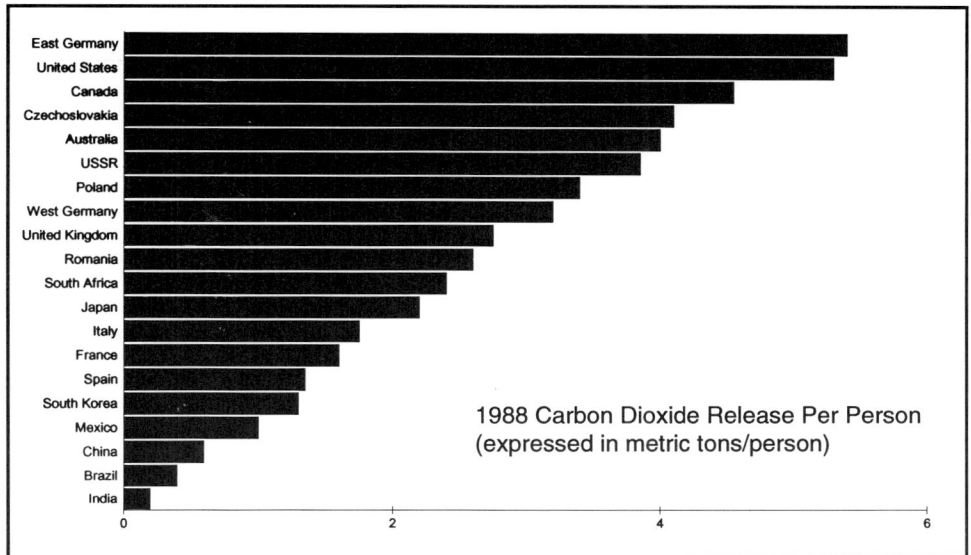

FIGURE 4   *When carbon dioxide releases are stated as releases per person, East Germany emits more than the United States. However, this does not take the amount of industrialization into account.*

worst culprit in 1988. Although the U.S. is not the worst polluter on a per-capita basis, it remains the second-worst emitter. In fact, the U.S. was virtually tied with East Germany.

The statistics can be twisted one other way—by stating emission in terms of how much a country produces. Obviously, countries that manufacture a great quantity of goods use more carbon dioxide because they use more energy. Figure 5 on the following page shows the relationship between production and emission. How much carbon dioxide countries release in producing the same amount of goods tells us who uses energy most efficiently. In these terms, the United States appears to be in the middle of the pack.

The most efficient countries in terms of carbon dioxide emissions per $1,000 of gross national product are France, Japan, and Germany. A large portion of the power produced in France and Japan comes from nuclear reactors—a much

larger fraction than in the United States. Nuclear reactors release no carbon dioxide. Still, those countries near the top of the efficiency list are to be congratulated.

Non-industrialized countries produce few goods with the carbon dioxide they emit. Some might feel they should leave the production to more efficient countries, but they would have no way of paying for the goods they must import. Perhaps the industrialized nations should spend their money helping poorer countries improve their energy efficiency, not in helping them produce more goods. Such a policy would win few friends for the industrialized nations, however, because the poorer countries want and need to produce more goods for their people. Most are too economically stressed at the moment to worry about what harm global heating might cause them in a few decades. Many nations resent being told to increase their energy-efficiency, because their best chance of making economic progress

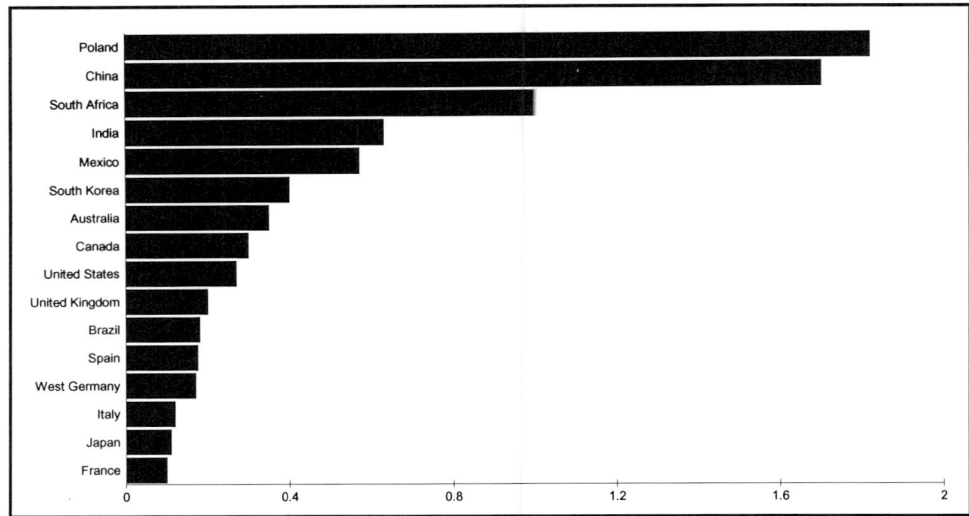

FIGURE 5 *Stating carbon dioxide releases in terms of production makes the United States look much better. Still, seven countries produce more gross national product per ton of carbon dioxide emitted than the United States. Based on production, all of the highly industrialized and rich countries emit at lower levels than the non-industrialized countries.*

involves producing as much as they can in any way that they can.

Even with the most efficient imaginable means of production, manufacturing enough goods so that all people of the world could live in a manner comparable to the lifestyles prevalent in developed nations would quickly quadruple atmospheric carbon dioxide. Stated another way, those of us in industrialized nations want to keep living like we do, but we know it means disaster if the rest of the world starts living the same way.

We can gain some perspective by considering a parallel thought. Norman Newell of the American Museum of Natural History and Leslie Marcu of City University of New York have made a startling suggestion regarding how we might measure population. They suggested counting world population by using atmospheric concentrations of carbon dioxide. Their justification is provided in Figure 6, which shows recent carbon dioxide and population levels. The two curves fall almost exactly in sync with each other.

This finding does not mean that carbon dioxide levels and population have a direct cause and effect relationship. Some people deliberately choose to consume less energy and benefit less from the "good life." In discussions of who is releasing too much carbon dioxide, they raise difficult questions: How well should we be allowed to live? How many material goods should we be allowed to have before we are living too well?

Cutting back on what we consume can, of course, reduce both energy consumption and carbon dioxide emission. However, for those who live in undeveloped countries, cutbacks could mean failure to survive. Only affluent countries can afford to cut back. Figure 5, carbon dioxide emission per $1,000 of gross national product, spurs a question of conscience for many people: Once we have achieved a certain level of energy efficiency, how well we should live?

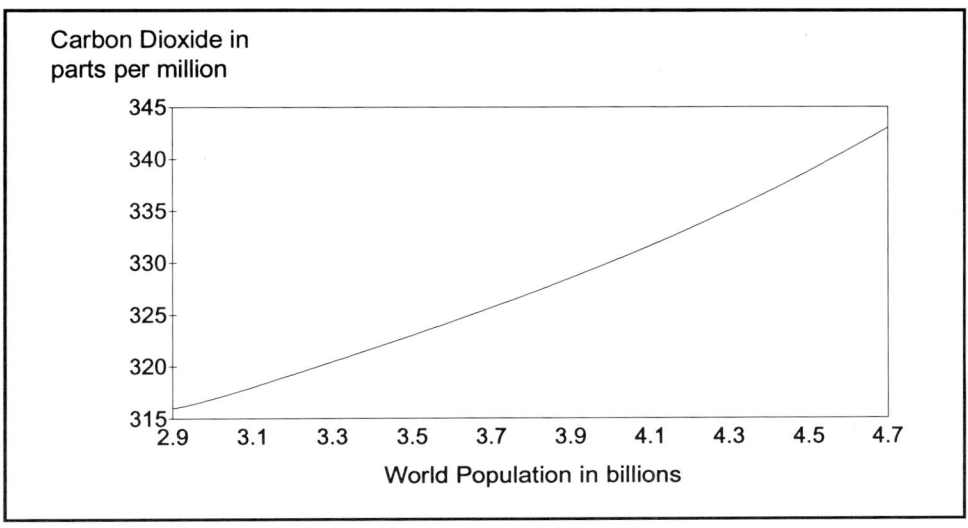

FIGURE 6 *Carbon dioxide concentrations and world population go up together, but this does not mean that each new person causes the same amount of releases. Most of the population increase comes from poor countries, where citizens cannot afford the consumption that leads to large releases. Most of the releases are coming from rich countries with relatively small population increases.*

CARBON DIOXIDE'S SHARE IN GREENHOUSE HEATING

Carbon dioxide has a bad reputation as the singular cause of global warming. It does in fact cause greenhouse heating, but looking at how much it causes, how much other gases cause, and how science works illustrates the complexity of the matter.

The current concern about greenhouse warming accelerated during the hot summer of 1988, when James E. Hansen testified before a Congressional committee. The hearing had little to do with greenhouse heating, but he responded to a question by saying, "It's just a logical conclusion that the greenhouse is here."

Hansen is director of NASA's Goddard Institute for Space Study and is heavily into climatic modeling. He could not have picked a more effective way to propel himself and his models into the forefront of public interest in science. Since giving that testimony, Hansen has been one of the most active investigators of greenhouse heating as a world problem. As global temperatures rise, so does his budget. Credit him with carefully spreading out the blame for the heating. He blames carbon dioxide, but he indicts other gases, too.

Figure 7, provided by Hansen, shows the expected greenhouse heating during two periods of time and the amounts that are created by different gases. In spite of the fact that concentrations continue to increase, carbon dioxide spurs a smaller fraction of the total greenhouse heating now than in the earlier period.

Other strange things emerge from this curve. Concentrations of methane shot up even faster than concentrations of carbon dioxide, but according to this curve, methane has decreased in importance. Carbon dioxide "forcing" (a term that here means driving up temperatures through global warming) increased, as you would expect from increased concentrations, but

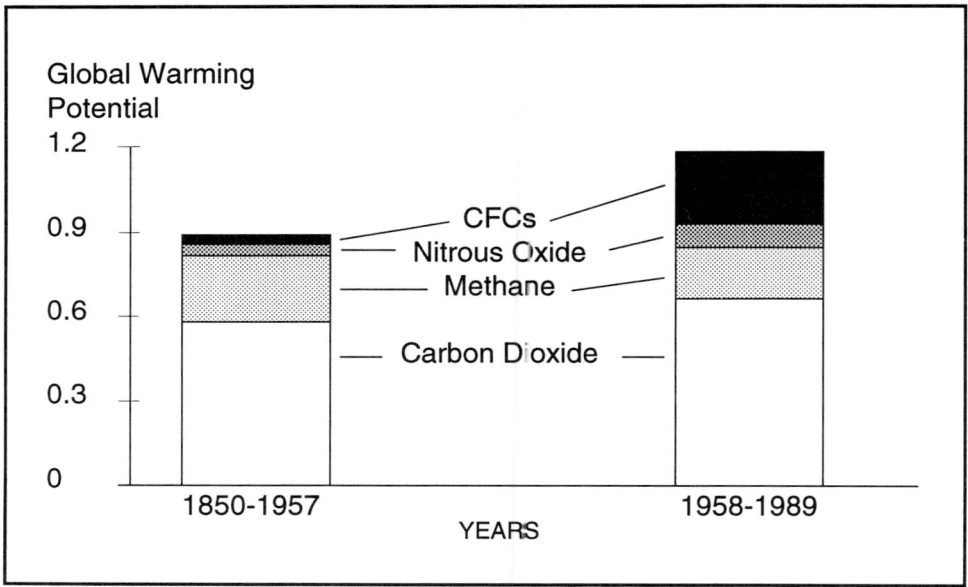

FIGURE 7   *Increases in greenhouse gases cause heat to be trapped, representing a warming potential. The increase in warming potential in the 107 years from 1850 to 1957 almost matched the increase in 31 years shown on the right. The relative contribution of gases changed, however. CFCs and nitrous oxide became much more important during the latter period.*

it provided a smaller fraction of the total forcing.

Here is a surprise. The air contained more methane in the recent period, but forcing by methane decreased. That decrease was due to nitrogen dioxide, which blocks radiation at the same wavelengths as methane. Nitrogen dioxide simply assumed some blocking that methane had been doing. CFCs were the real growth item in greenhouse forcing. Most CFC releases occurred after 1957. Before then, people could not afford so much air conditioning.

One point illustrated by Figure 7 is this: Carbon dioxide gets the bad press, but even with water vapor emissions omitted, carbon dioxide now accounts for just a little more than half of the problem. It seems that press and television reporters learn a term like "carbon dioxide" and never bother learning much more. They keep beating the same old horse.

But others beyond the media are responsible for the lack of public understanding. Scientists themselves often tout a popular catch-phrase without understanding what it really means. They become so specialized that their understanding of the subject as a whole barely surpasses the general public's understanding. For example, the people who do climatic modeling know very little about absorption by different gases. They trust someone else to tell them how much gases really absorb. The researcher giving the numbers on absorption to the climate modeler probably does radiation modeling and has never even seen anyone make an absorption measurement. Knowledge keeps branching out this way. In some cases, no single person develops a thorough understanding of the entire process. People doing measurements of absorption select what they measure from what the climate modelers say is important.

One result is that specialized scientists try to educate the public by saying things that sound good, but which turn out to be nonsense. Here is one example. Figure 7 appeared in the book *Global Warning . . . Global Warming* by Melvin A. Benarde, a professor of environmental studies at Temple University.

After presenting the figure, Benarde tried to punch it up by saying:

"As the figure shows, the freons have contributed substantially to this increased warmth. Perhaps as disturbing is the fact that, even if their production and release stopped tomorrow, their atmospheric concentrations would continue to increase because of their inordinate atmospheric residence times—on the order of 60 to 100 years."

This information is wrong. It illustrates what happens when an expert tries to talk outside of his field. Benarde claims that, because freons stay around a long time, their concentrations will keep increasing. Not so. Once we stop releasing freons, their concentrations cannot increase.

Concentrations of chlorine, but not CFC concentrations, in the upper atmosphere might continue to increase for 60 to 100 years after CFC releases stop, but not because of "inordinate atmospheric residence times." That is simply how long it takes CFCs to migrate from the lower atmosphere to the upper atmosphere, then to have their chlorine released by solar ultraviolet (thus decreasing their concentrations).

This chlorine is central to the destruction of ozone, but the figure has nothing to do with destruction of ozone by chlorine compounds. The figure deals with warming by CFCs—only some of which are freon. Once the solar ultraviolet breaks down a CFC molecule, global forcing by

that CFC molecule ceases. The CFC molecule is history; it no longer exists. Beyond that, CFC molecules are more effective as absorbers in the lower atmosphere than they are in the upper atmosphere. Stop releasing CFCs and their effects on ozone may continue to increase for years, but their effects on global warming will immediately start decreasing.

In the argument over global warming, this instance of misspeaking is unimportant. However, such misinformation seems to amount to a scientific epidemic. Many scientists try to stretch their work far enough to connect it with environmental concerns, even though they may not understand the environmental concerns. And they rely on the silence of their colleagues, assuming that no one will point out how wrong or nebulous the alleged environmental connection really is. Connecting one's research with a topic of vital public interest cannot hurt, and may help the scientist obtain future research grants.

What's the problem? Many of these alleged scientific "connections" confuse, mislead, or misinform all of us. They don't help us spend our money wisely; nor do they help us solve the real problems at hand.

CARBON DIOXIDE AS AIRBORNE FERTILIZER

Some people see increased carbon dioxide as the door to a brighter future. They say that warming some of the cooler regions of the world would mean extending the growing season, resulting in more food for more people. They point out that global circulation models predict generally greater rainfall throughout the world, but less rainfall in some regions like the southern United States. More rain produces more crops, which means more food for more people. The results sound good, even without factoring in the positive effect carbon dioxide has on plant growth.

Plants grow by using sunlight to convert carbon dioxide into vegetative growth—whether the growth is fruit, stalk, or leaf. Increased concentration of carbon dioxide causes plants to grow faster and to produce larger crops. Such plants also make better use of water, which means they are more drought-tolerant. There is a downside, of course, although it has nothing to do with plant growth. The downside is the pain involved in adapting to a significantly altered world.

Acting as an airborne fertilizer, carbon dioxide might do more than generate a greener world. It might also limit the increase in carbon dioxide to less than the doubled level that is assumed for model computations. Gregg Marland at Oak Ridge National Laboratory thought that increased forest growth might resolve the carbon dioxide problem. He did more than talk about the idea. He went to the trouble of calculating whether and how such growth might help.

Marland first investigated whether enough new forests could be planted to absorb the excess carbon dioxide we release into the atmosphere. The answer was yes, but not easily. To absorb the excess would require planting an area the size of Australia with a fast-growing variety of tree, like the sycamore. Sufficient land exists for such forests, but persuading countries to dedicate that amount of land seems highly unlikely. So far, most countries have resisted efforts to persuade them to stop cutting the forests they already have.

Some investigators have argued that harvesting old forests and replanting the land with young trees could reduce carbon dioxide. New forests grow faster, they say, and might be expected to incorporate more carbon as they grow. That argument considers nothing but growth. However,

something must become of the wood in the old forests. In normal usage, almost half the harvested wood becomes paper or wood chips. These products are soon burned or decayed, and their carbon returns to the atmosphere. The result of cutting old trees to plant young ones is a net increase in atmospheric carbon dioxide, because more wood from the cut forests gets converted back into carbon dioxide than the new growth can absorb.

After concluding that countries would not commit enough new forest land to overcome the increase in atmospheric CO_2, Marland suggested another approach. Instead of planting more forests, why not increase the growth rate of existing forests? This could be done through irrigation, fertilization, and pest and fire control. He found that the growth rate of the world's forests would have to double in order to absorb the excess carbon dioxide now dumped into the atmosphere.

Marland's conclusion was a glum one: "Looking to forests to solve the carbon dioxide problem is unrealistic."

Sherwood Idso at the U.S. Water Conservation Laboratory in Phoenix, Arizona, pounced on Marland's finding that doubling the forest growth rates would capture all the excess carbon dioxide. Idso claimed that the airborne fertilizer effect of the carbon dioxide might accomplish the doubling. Most biologists believed the airborne fertilizer effect was too small to double growth rates, and they were correct. Tests with crops and other small plants indicate that doubling the concentration of carbon dioxide increases growth rates by perhaps one-third.

Idso realized that photosynthesis operates differently in woody plants than in plants previously tested in atmospheres with enriched carbon dioxide. He grew sour orange trees in doubled carbon dioxide atmospheres. Trees in the enriched

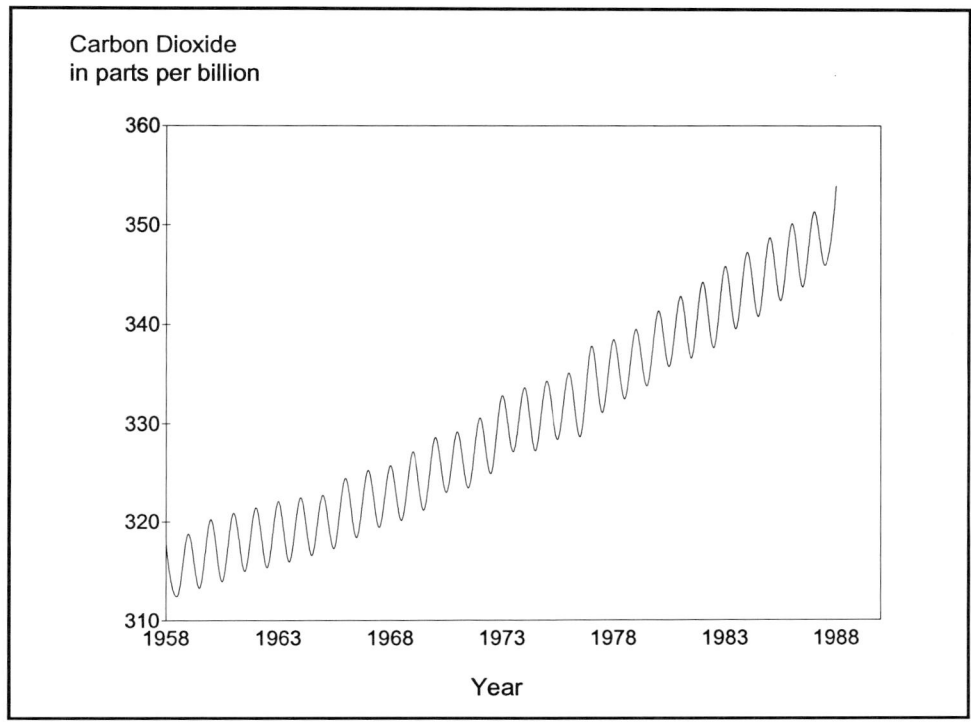

Carbon Dioxide
in parts per billion

FIGURE 8 *Carbon dioxide concentrations in the atmosphere appear to be ever-increasing, but there are seasonal variations. As spring comes to the northern hemisphere, plants grow and remove carbon dioxide from the air. Computer analysis of this annual cycle shows that it becomes larger each year. This may happen because plants grow more when the air contains more carbon dioxide.*

atmosphere stored 2.8 times more carbon as biomass than controls that were grown in normal air. Idso emphasizes that this factor of 2.8 has to do with stored mass— not the factor by which photosynthesis increases.

At first glance, photosynthesis and conversion to biomass seem equivalent. More work needs to be done on this question, because Idso's results differ from results obtained by others who grew trees in the same sort of atmosphere. The results are not directly comparable, but a report says sugar maples increase photosynthetic uptake by only 20 percent and show no increase in growth rate. Beech trees increase uptake only half as much as sugar

maples, but they *do* show an increased growth rate. Even experts cannot determine what all these results really mean. They cannot agree about carbon dioxide's airborne fertilizer effect.

Idso did not stop after he found that an enriched atmosphere caused the trees to convert 2.8 times as much carbon to biomass. If he had, his result would simply reflect one particular tree species that responded differently from others. He went on to generalize his result by using the very data that documents the history of carbon dioxide increases.

The longest and best set of precision carbon dioxide measurements are those initiated by C.D. Keeling on Mauna Loa

(Hawaii), shown in Figure 8 on the preceding page. Annual growth cycles cause yearly peaks and valleys in the carbon dioxide concentration. The reduced concentrations occur in the northern hemisphere spring. As plants awaken from dormancy, they withdraw large quantities of carbon dioxide from the atmosphere, then release it in the fall.

Computer analysis of the peak and valley data in Figure 8 yields information that the eye cannot discern. The annual cycle gets larger each year. Differences between the peaks and valleys of the concentration yield a smooth increase when mathematically analyzed. Valley to peak amplitudes grow a half-percent larger each year. This increase in amplitude comes from increased plant growth, increased carbon dioxide withdrawal in the spring, and larger CO_2 releases in the fall. Thus, some fertilizer effect must exist.

By realizing that the ever-growing cycle swings were due to increased plant growth, Idso could calculate how worldwide growth responds to increased atmospheric carbon dioxide. In other words, he used the increase in the annual cycle to determine how the airborne fertilizer increases growth. Growth of land-based vegetation accounts for 90 percent of the annual cycle amplitude, and trees provide 75 percent of the photosynthesis. These values and Marland's finding indicating that a doubled growth rate of forests would absorb enough carbon dioxide to cut off its increase were enough for Idso. He used the data to arrive at a beautiful answer.

When he put these pieces of data into an equation, Idso found that increased growth rates would trigger absorption of all the carbon dioxide increase before the atmospheric concentration had doubled. All the current model computations that project global heating in the next century assume double the current concentration of carbon dioxide. Idso's result indicates that a doubled concentration could only come about if emissions increase. At the present rate of carbon dioxide dumping, a carbon dioxide increase of 50 percent is the maximum that could occur. That 50 percent increase in carbon dioxide would cause plants to grow fast enough to absorb all the excess carbon dioxide now released into the atmosphere.

Recall the conclusion of Tans, Fung, and Takahashi—that the major sink for carbon dioxide was situated on land, not in the ocean. Their result is consistent with Idso's. The trio says most of the carbon dioxide remains on land, but they don't say where it goes. Idso says plants are absorbing more and more carbon dioxide, but he says nothing about the oceans. The two sets of results fit together. Again, both sets of results remain unproven, but no one has disproven them yet, either.

Strangely enough, the world yawned at this news. No one has poked holes in Idso's result or shouted, "Hurray, the world is saved!" Perhaps no one sees this result as salvation because people are rapidly cutting the forests that Idso counts on to absorb the excess carbon dioxide. Cutting the forests means they cannot grow more rapidly, and cannot counteract the increase in atmospheric carbon dioxide.

We all like to arrive at the real truth and the final answer. This chapter leaves the reader uncertain about just what happens (and what will happen) to carbon dioxide. That is disappointing, but it also reflects the true state of our knowledge. No one knows. In fact, in my experience, only those with an axe to grind or a profound bias insist that they have the ultimate answer.

7

Methane, Greenhouse Gases, and Friends

You may have to give up beef as well as your car.

Water is the most important greenhouse gas, followed by carbon dioxide, but methane, several oxides of nitrogen, ozone, and even the CFCs that destroy ozone contribute to global warming. The phrase "greenhouse gases" demands a careful definition, because some molecules that are not greenhouse gases cause global warming. These molecules do not affect outgoing infrared radiation as gases that cause greenhouse warming do, but they do cause atmospheric heating by absorbing visible or ultraviolet radiation from the sun. Absorbing solar radiation may affect temperatures, but the greenhouse effect does not refer to such absorptions.

Still other molecules have no important absorptions or emissions at either the solar or infrared wavelengths. Nonetheless, these molecules produce or destroy green-house gases. Some discussion of these molecules can be helpful in understanding where greenhouse molecules come from and what limits their concentration.

First, a recap of how molecules contribute to greenhouse heating. The sun causes little direct heating of the air, because air absorbs only a small fraction of the sun's heat. Most of the sunlight gets absorbed by the ground and by plants. This warms them up, and they transfer their warmth to the air that blows along the surface, thereby cooling off the surface. Solar energy absorbed at the surface either heats the air or the surface re-emits the heat as infrared radiation. Greenhouse heating occurs when molecules of air absorb part of the infrared energy radiated by the surface or by the air itself, thus preventing it from escaping into space. Without greenhouse gases in the atmosphere to absorb the radiation, it would escape into space, leaving the world a little cooler.

After a molecule absorbs radiation, there is little it can do with the absorbed energy other than re-emitting it. If nature were arranged so the molecules re-emitted all their radiation upward, no greenhouse heating would occur. Unfortunately, the molecules emit radiation in all directions. Some goes back down, to be reabsorbed by the earth. You might say the atmosphere shines back onto the ground. So greenhouse gases absorb radiation from the surface, then re-emit part of it back to the surface. As a result less radiation escapes to space, and the surface becomes hotter.

Molecules at colder temperatures than the surface return less energy to the surface. Thus, the effectiveness of an absorbing gas depends on where the gas is located in the atmosphere. Some gases, like ozone, are more concentrated at high altitudes. Others are concentrated at low altitudes, like water vapor. Still others, like carbon dioxide, form a constant fraction of the atmosphere at all altitudes.

If the gas is a weak absorber, doubling its concentration level doubles the absorption, and doubles that gas's contribution to greenhouse heating, also known as "radiative forcing." With large concentrations of the gas, the absorption is almost total, so adding more gas causes little additional absorption. In this case, doubling the amount of gas would not double the molecule's radiative forcing.

The concentration of methane has doubled since pre-industrial times, but its contribution to global warming has not doubled. Likewise, when two different gases absorb at the same wavelength—as methane and nitrous oxide do—absorption increases, diminishing the greenhouse effectiveness of both gases.

WHO'S WHO OF THE GREENHOUSE

As we've seen, the atmosphere contains an abundance of carbon dioxide. At such high concentrations, absorption becomes almost total, and radiative forcing falls off. A few more molecules of carbon dioxide cause less radiative forcing than earlier molecules caused. Still, carbon dioxide is the main gas that triggers the greenhouse problem. To illustrate: If one molecule of the minor gas absorbs 1,000 times as much as one molecule of carbon dioxide, the minor gas still has little net effect if the atmosphere only contains a few of its molecules.

One of the easiest ways to describe the effect of gases on greenhouse heating is to talk about changes that have occurred since the Industrial Revolution. The world was changing even before then, still warming from the little ice age. However, it is traditional to use the climate around 1860 as a base level, because many experts consider earlier temperature records unreliable. Few meteorological stations existed before the Industrial Revolution, and these were concentrated in Europe, the British possessions, and the eastern United States. Such spotty temperatures represented particular regions, but they did not provide an accurate picture of temperatures on a worldwide basis. Some critics claim that even the stations within modern meteorological networks are too spread out to yield proper worldwide averages.

In any event, temperature increases are not a proper measure of greenhouse contributions. Temperature increases do not work because, during some decades, temperatures have decreased. Instead of talking in terms of temperatures, the Intergovernmental Panel on Climate Change (IPCC) used the term "radiative forcing," which allowed the panel to address changes even during the times when temperatures decreased. Radiative forcing simply measures warming in terms of how much radiation a gas blocks from escaping into space. Interestingly, the panel picked 1765 as its starting date—well before the existence of reliable weather data. About the best feel the panel could have for what radiative forcing means is that the larger the forcing number, the greater the trapping of heat. No one has ever measured radiative forcing; it is strictly a calculated number based on concentrations of the various greenhouse gases.

Figure 1 shows the radiative forcing caused by different gases back to the days of the Industrial Revolution. Since the forcing is greatest during recent times, the intervals on the right are shorter, and the graph takes that into account. The who's who of the greenhouse gases includes all the molecules given in the graph and one more—ozone. Although not included, it is an important greenhouse gas. It may seem strange that the figure includes stratospheric

water vapor, but not water vapor in the lower atmosphere. This is because stratospheric water vapor forms from the decomposition of a greenhouse gas, methane, while water in the lower atmosphere simply evaporates from the oceans and soils.

No one really knows how much ozone the global atmosphere contained before about 1960. Surprisingly, scientists know the concentrations of the other gases fairly well for the listed times—perhaps better than they know the temperatures. Gas concentration figures come from measurements of air trapped in samples of polar ice. Many gases can remain trapped in the ice, like food in a deep freeze, without changing or losing their identity. Ozone, however, is not one of those gases. It breaks down when trapped in ice and becomes another oxygen compound. Just as importantly, most of the ozone remains

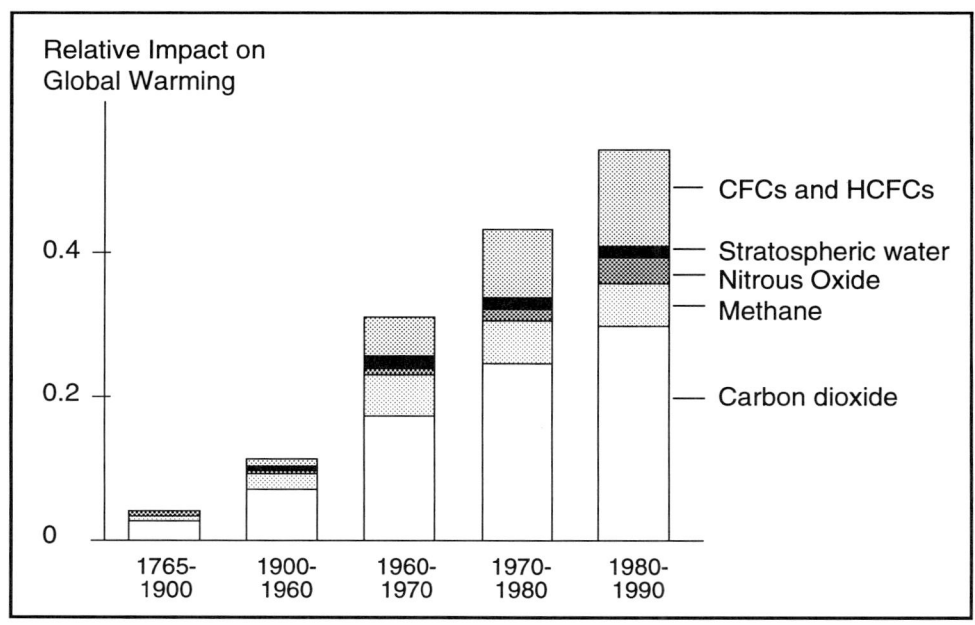

FIGURE 1 *Increases in greenhouse heating come from gases other than carbon dioxide. Little is known about what caused the increase in nitrous oxide, and debate continues about the increase in methane.*

uncaptured even in the upper atmosphere.

Polar ice can only capture lower atmospheric gases. There is no way that polar ice could have trapped stratospheric water vapor from the upper atmosphere. Even if it had, the vapor would simply have turned into more ice. The IPCC thought that water vapor was too important to omit, so they based the concentration of water vapor on the concentration of methane in the lower atmosphere. As methane was oxidized by solar ultraviolet radiation, it produced the stratospheric water vapor. As methane increased, the IPCC assumed that stratospheric water vapor also increased, and guessed how much might have been present in the upper atmosphere.

Only 51 percent of radiative forcing is due to carbon dioxide. Methane accounts for 17 percent of the forcing; the stratospheric water vapor for six percent; and nitrous oxide for four percent. The sections that follow discuss these gases, where they come from, and what removes them from the atmosphere.

Earlier chapters dealing with the ozone hole discussed the CFCs and HCFCs in detail. We should not belittle their importance as greenhouse gases; they account for 12 percent of the forcing. The quantity of CFCs in the atmosphere is not as large as the quantity of other greenhouse gases, but each pound of CFCs produces some 4,000 times the heating of a pound of carbon dioxide.

International agreements to cut back on CFC production provide no forgiveness for the massive quantities already released. In spite of cutbacks, large concentrations of those chemicals will be floating around in the atmosphere 50 years from now. Replacements for these ozone destroyers may not affect ozone, but the replacements appear to be even stronger greenhouse gases than the CFCs. The shorter lifespans of the CFC replacements minimize their

effect on the climate, but the old CFCs disappear slowly from the atmosphere. Until the old chemicals disappear, both the old CFCs and their replacements will be in the air, acting as greenhouse gases. During that period, the combination of old and new will significantly increase warming. We can expect that things will get worse before they get better.

Other gases affect the life cycle of the greenhouse gases. These gases in and of themselves contribute almost nothing to the radiative forcing; nonetheless, they affect forcing through their ability to change the lifespan of other gases.

The most important of the molecules that create and destroy greenhouse gases is the combination of oxygen and hydrogen, OH, sometimes called the "hydroxyl radical." We know little about how much OH the atmosphere contains, how it varies with location, or when the level decreases. Our ignorance is not due to a lack of interest—information is hard to obtain, because OH is so reactive. Researchers have tried to capture it, but OH interacts with whatever substance is used to capture it before the amount can be analyzed. A new laser measuring system measures the hydroxyl concentration by passing a laser beam through open air. The new technique may tell us more about this mysterious but important chemical.

Another gas that contributes little warming of its own is sulfur dioxide. It condenses into aerosols, which scatter light and affect cloud formation. Nitrogen compounds also condense into aerosols that scatter light and cause haze.

ATMOSPHERIC METHANE

Looking at the concentrations of methane and the temperature record over 160,000 years, shown in Figure 2, is either impressive or scary. The temperature and methane concentration curves look quite

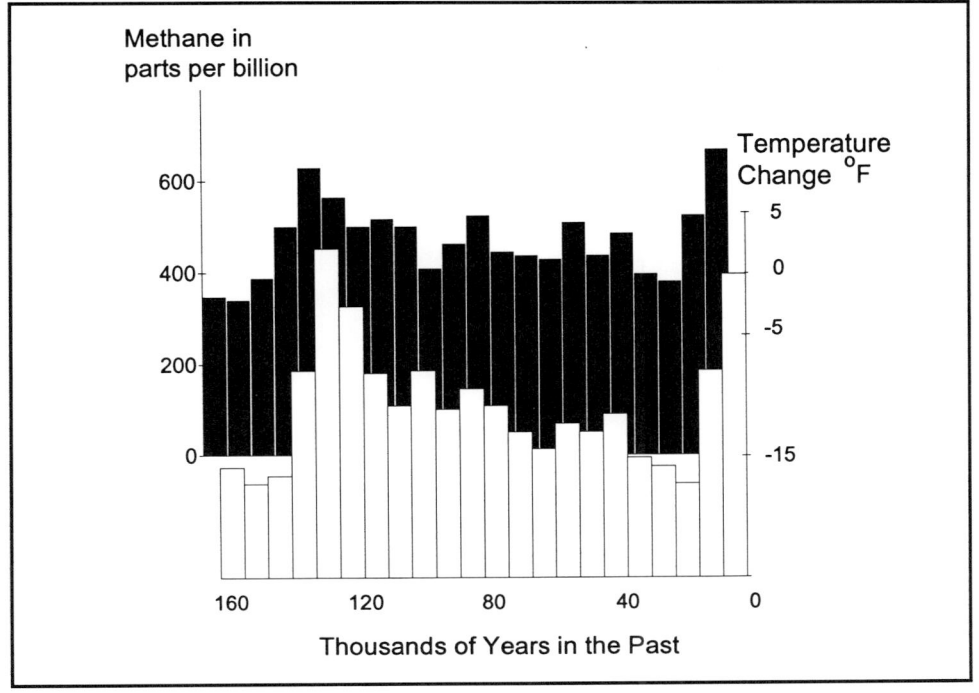

FIGURE 2 *Methane has been steadily increasing for 300 years. Its concentration has almost tripled, while concentrations of carbon dioxide have only increased by about 30 percent.*

similar—too similar to be a product of chance. This is a case of, "Which came first, the chicken or the egg?"

The first thought is that the increase in methane might have caused the heating of the atmosphere. The second thought is that the heating of the atmosphere might have caused more methane production.

Perhaps warmer weather and increased carbon dioxide caused more vegetation to grow. As vegetation died, part of it turned into methane. Which came first need not be an either-or situation. The methane from decaying vegetation contributes to greenhouse heating. In turn, the warmer climate grows more vegetation, which decays and produces more methane. The opposite effect could occur as the world cools. But what causes heating and cooling trends to reverse? What triggers ice ages

and brings them to an end? These remain open questions.

Right now, the atmospheric concentration of methane is 1,720 parts per billion—an amount too large to fit on the graph. That concentration represents more than twice as much methane as at any time in the last 160,000 years. The rapid increase during the last 100 years can be seen in Figure 3.

Methane is currently increasing at 0.9 percent per year. Since pre-industrial times, concentrations of carbon dioxide have only increased 30 percent, but concentrations of methane have doubled, representing a 100 percent increase. Although methane has shown a higher percentage rate of increase, there is more carbon dioxide in the atmosphere, and the weight of the carbon dioxide is increasing more rapidly than the

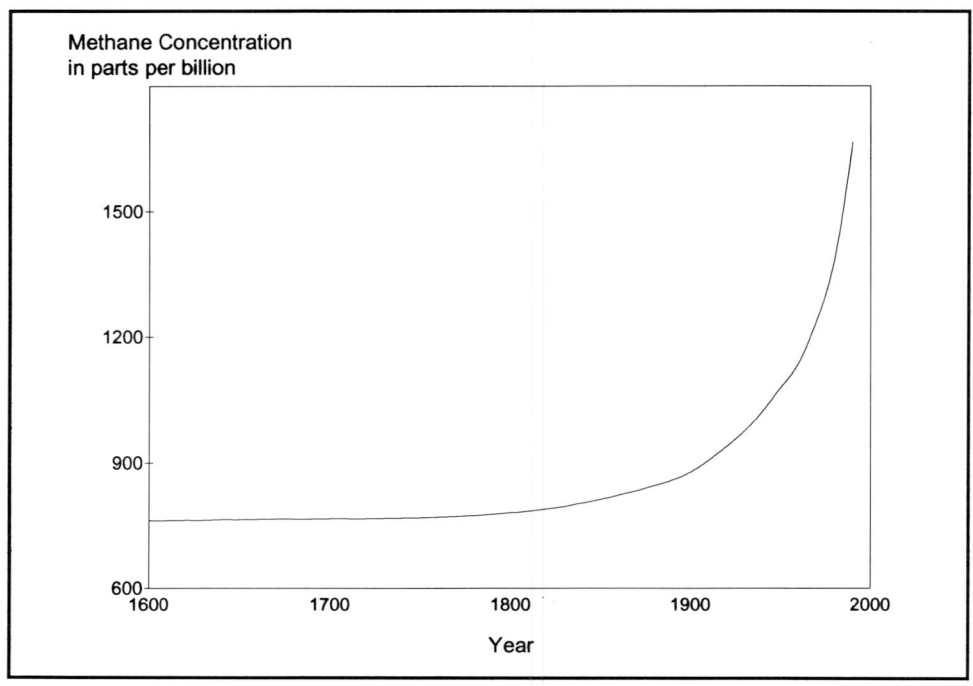

FIGURE 3 *Increases in global temperatures and methane concentrations go hand in hand. This does not mean the increases in methane caused temperatures to increase. Perhaps warmer temperatures caused increased methane production though increased plant growth.*

weight of methane. In fact, at current release rates, the mass of carbon dioxide in the atmosphere increases 80 times faster than the mass of methane.

Methane is nearly inert; it scarcely interacts with other atmospheric gases. The one gas with which it does interact is OH, a reaction that converts methane into carbon dioxide and water vapor. In the lower atmosphere, the water vapor joins the first cloud it encounters and falls back to the earth as rain, leaving only the carbon dioxide in the air.

Oxidation of methane into water vapor is more important in the upper atmosphere, because it is one of the few ways that water vapor enters the stratosphere. The stratosphere lays just above the coldest part of the atmosphere, the tropopause. Below the tropopause, air gets colder at higher alti-

tudes. Above the tropopause, the air gets warmer at higher altitudes, though never anywhere near as warm as surface temperatures. The "pause" portion of the word "tropopause" is there because the temperature change pauses—neither increasing nor decreasing for some span of altitude.

Almost no water escapes from the moist troposphere upward into the stratosphere. Water vapor is frozen out of the air as it diffuses through the frigid temperatures in the tropopause, and this deprives the upper atmosphere of the most effective greenhouse gas—water vapor. Here, methane plays a role. It does not freeze out at the tropopause; instead, it travels into the upper atmosphere and reacts with hydroxyl to become water vapor and carbon dioxide.

As the graph of radiative forcing indicated, stratospheric water vapor produced from methane provides six percent of the overall radiative forcing. Stratospheric water vapor also forms the ice crystals that are so important in the destruction of ozone. As a result, increases in methane may be bad for the ozone and may contribute to global warming as well.

Methane is formed by joining one carbon atom with four hydrogen atoms. It is the simplest of the organic molecules, and its family includes methane, ethane, butane, and pentane. These gases fall into the hydrocarbon category, because they contain only hydrogen and carbon atoms. At one time, chemists thought only the miracle of living organisms could produce such molecules—thus they came up with the name "organic chemicals." Natural gas used for heating is largely methane, but it generally contains other members of the methane hydrocarbon family. Methane itself is odorless. The odor of natural gas is a safety measure—a chemical added so people can smell gas leaks.

Bacteria that eat away at decaying materials produce most of the methane in our atmosphere. Only things that were once alive can decay—substances like wood, plants, and animals. Because of this, many people think of methane as a gas that always emerges from some life-related process. In fact, life forms are usually involved here on earth, but most of the methane in the universe has nothing to do with bacteria or life forms. If organic matter decomposes without oxygen, methane forms. If oxygen were present, it would oxidize the organic matter's carbon into carbon dioxide. Thus, photosynthesis produces organic plant matter. After digestion by some creature, the plant matter ends up as methane or carbon dioxide. Both are greenhouse gases.

Methane exists in other parts of our solar system and elsewhere in the universe, but the earth is the only planet on which methane suggests living processes. A stellar nebula contains hydrogen and carbon that often react to form gaseous methane. As the nebula becomes a sun, oxygen burns the methane away. The atmospheres of several other planets contain methane. Jupiter's atmosphere, for example, contains massive amounts of methane, but Jupiter never supported life forms to generate the methane.

Since methane occurs on lifeless planets, one scientist insisted that natural gas—methane—forms deep inside the earth through a process unrelated to petroleum formation. He argued that although oil wells always contain some natural gas, natural gas fields sometimes contain no oil. To test his theory of non-petroleum methane generation, he convinced a group to drill a deep test well. The well found no significant gas deposits, but the results were inconclusive. Theories such as his are well-nigh impossible to disprove. One successful test might verify the theory, but bad luck could explain any number of failures to find methane.

Most geologists believe that natural gas forms along with oil from trapped organic matter. This strongly-held belief, combined with the inconclusive test well experiment, have effectively discredited the non-organic natural gas theory. Some hoped that the theory would be verified, because verification might indicate that the earth itself holds tremendous natural gas energy reserves.

Advertisements promote natural gas, which is primarily methane, as an energy source that produces less air pollution than other combustion processes. Although this is correct in terms of pollution, it has no bearing on the generation of greenhouse gases. Burning methane produces carbon dioxide, just like any other combustion process. The energy obtained from each molecule of carbon dioxide differs from fuel to fuel. However, natural gas and all other fossil fuels release carbon dioxide when burned.

Articles that attack cows for generating methane never fail to compare the greenhouse effectiveness of carbon dioxide and methane. Readers may encounter a whole catalog of such numbers, not all of which agree with each other. Figures cited include the amount of radiative forcing per molecule of methane and carbon dioxide— the methane molecule yields 21 times as much heating.

A molecule of methane weighs less than a molecule of carbon dioxide. If stated in terms of weight instead of on a per-molecule basis, a pound of methane produces 58 times as much forcing as a pound of carbon dioxide. However, the matter is not quite so simple.

Methane lasts for only about ten years in the atmosphere, while carbon dioxide has an effective life of 50 to 100 years. (Remember, also, that methane eventually oxidizes and ends up as carbon dioxide.) Lifetime is a consideration. Over 20 years, a pound of methane causes 63 times as much heating as a pound of carbon dioxide, but over 100 years the methane causes only 21 times as much heating. During part of this time period, radiative forcing is less because the methane has already been converted to carbon dioxide. That conversion accounts for the lowered effectiveness of methane over long time periods. These numbers are culled from different sections of the IPCC report. The statements that a pound of methane is 63 times worse over 20 years but only 58 times as bad if considered by weight disagree. Both cannot be correct. Many people wrote sections of the IPCC report, and all apparently opted to report their favored values. Just because a scientist or a journalist uses a number does not mean the number is correct.

WHAT MAKES METHANE

Methane generation and carbon dioxide releases are virtual opposites. One sounds natural, the other industrial. Carbon dioxide comes from power plants, automobiles, and other activities associated with industry. There are natural sources of carbon dioxide, like decaying plants, but they are natural recycling operations. Plants absorb carbon dioxide from the air as they grow and return it as they decay. Plant growth and decay are repeated cycles that cause no net increase in atmospheric carbon dioxide. The carbon dioxide increases come from new activities—from machines and energy generation. When presenting data on the growth of carbon dioxide, graphs always start with the Industrial Revolution.

Methane does not fall in the same camp. It is generated by the breakdown of living matter—primarily vegetation—so it seems to comes from wholesome sources like crops, animals, and other activities that have nothing to do with industry. A couple of industrial sources exist, but these are minor compared to the natural sources.

Plant matter has always broken down, so why should methane levels be increasing? We produce more food than we used to, and this essentially explains why more methane gets into the atmosphere now than was the case in pre-industrial times. The Industrial Revolution gave us the machines, chemicals, and techniques for producing more food. More food meant populations could increase without starving. The world population increased, and methane production increased along with the population.

We could be producing a larger portion of our food in ways that cause no methane release, but only if people changed their eating habits. Some suggest that we should give up beef, but rice paddies also produce methane. Giving up rice would not be nearly so simple, because it is a food staple for many of the underdeveloped nations. Methane release raises environmental questions about things like beef and rice—questions that environmentalists often try to ignore.

When substances containing carbon break down, the carbon can end up in carbon dioxide or broken down into methane. Any of several different breakdown paths may be followed, but in the end the reactions produce either carbon dioxide or methane.

When oxygen is available, the carbon in plant and animal matter eventually oxidizes into carbon dioxide. Carbon dioxide is a sort of sandwich, with one carbon atom situated between two oxygen atoms. Organic matter contains little or no oxygen. Without oxygen, there is no way the carbon in organic compounds can become carbon dioxide. However, organic compounds do contain hydrogen. When no oxygen exists for chemical reactions and hydrogen is available, hydrogen bonds with the carbon atoms.

The bonding together of one carbon and four hydrogen atoms generates methane, and since organic matter contains both hydrogen and carbon, methane can form by using only the materials in plant and animal matter. Methane forms when organic materials break down without enough air to supply oxygen.

Vegetable matter that rots underwater is a common source of methane. Bubbles of methane often rise in swamp waters, which is why methane is also called swamp gas. When plants decay in swamp water no oxygen is available, so carbon dioxide cannot form. Inevitably, the hydrogen and carbon in the plant change their form and bond together—and out comes methane. But in a sense, the methane-forming reactions are only half-complete. The methane will eventually oxidize into carbon dioxide, which completes the reaction.

Many methane-producing reactions are driven by anaerobic bacteria (bacteria that live without oxygen). Such bacteria produce methane when they digest organic matter. Anaerobic digestion occurs underwater and in the digestive tracts of many animals.

WHO MAKES METHANE?

Table 1 reveals the major sources of

Source	ESTIMATED ANNUAL RELEASE (Millions of Tons)	RANGE (Millions of Tons)
Commercial operations		
Gas drilling, venting, transmission	45	25 – 50
Biomass burning	40	20 – 80
Landfills	40	20 – 70
Coal mining	35	19 – 50
Animals		
Intestinal fermentation	80	65 – 100
Termites	40	10 – 100
Under water and wet soils	?	?

TABLE 1 *The center column gives the IPCC s best estimate of sources and sinks of methane. The right column lists the possible range of values. As the values show, estimates have large possible errors.*

methane. The gas comes from three kinds of activities: Formation in commercial operations, formation in wet soils and underwater, and formation by animals.

Methane causes most coal mine explosions. The carbon in plant matter that was buried in the geologic past eventually formed deposits of coal. Some hydrogen from the plants escaped, and no oxygen was present, leaving only a complex, carbon-based mix that turned into coal. But not all of the hydrogen escaped or was captured in solid coal. Some reacted to form methane, which ended up trapped as gas pockets in the coal. As miners remove coal, they inadvertently open these pockets and spill methane into the rest of the mine. If a spark ignites the methane, it explodes. The spark initiates burning—the reaction that allows the carbon in methane and oxygen from the air to combine into carbon dioxide.

Explosions are nothing but burning that takes place at a fantastic rate. All the hydrogen in the methane combines with oxygen and ends up as water vapor, or H_2O. The carbon combines with oxygen, too. The result is carbon dioxide. Most of the energy released by burning results from the combining of hydrogen and oxygen to form water vapor. The formation of carbon dioxide releases only a fraction as much energy.

Animal matter trapped in past geological ages forms into oil and natural gas. The chemical changes in the animal matter are similar to those that occur in the formation of coal. Again, some carbon and hydrogen atoms from the trapped animal matter combined to form methane, i.e. natural gas. This became the gas that wells tap into to obtain the natural gas we use today.

Experts think that most of the natural gas released into the atmosphere comes from small leaks in the gas pipeline distribution system. Those experts can only guess at the amount of methane that

escapes in this manner. Each leak is small. Estimating how many leaks there are and the average size of the leaks is a process subject to large errors. Unfortunately, most estimates concerning methane sources represent only rough estimates, rather than measurements.

Table 1 includes methane clathrate destabilization due to commercial drilling as a process responsible for releasing some methane into the atmosphere. Clathrates are ice-like deposits of methane and water that exist in the permafrost of Alaska and Siberia. Warming temperatures and drilling operations have caused the methane in some of these deposits to evaporate into the atmosphere. At this time, the clathrates are not an important source of methane, but global warming could ultimately trigger the release of tremendous quantities of methane.

Rice grows in wet fields that are flooded by farmers during early growth stages. Repeated flooding turns rice paddies into major sources of methane. Actually, the methane is produced by bacteria at work on roots and buried straw from previous crops. If the fields were not wet or flooded, the buried plants would decay and form carbon dioxide, but the water keeps oxygen out of the decay process. Without oxygen, methane forms.

Rice production has increased as the world's population increased. Some increase came from improved plant varieties, some from the use of fertilizer to improve yields, and some from additional land converted to rice production. Without this increased production, masses of people would be driven to the edge of starvation. For example, India is now self-sufficient in terms of food production, but only because of increased rice production.

Few measurements exist, but cutting tropical rain forests may contribute more methane to the atmosphere. Cutting the forests turns the roots that formed part of living trees into dead wood—wood that decays. The heavy rainfall keeps the ground saturated, so the roots decay with little or no oxygen. Methane from the decay diffuses up and out of the soil. Without this accelerated decay, much of the root mass would become the organic material that forms topsoil. Soils in forests hold a large reservoir of carbon.

Regulations demand that sanitary landfills use liners or clay soil to prevent water from leaking through their bottoms. Leakage could carry landfill poisons into the water table. But in the absence of leakage, the waste just lays there in water and wet soil, conditions that cause decomposing trash to turn into methane.

Many people urge protection of wetlands, but wetlands contribute to the methane problem. Swamps are only one kind of wetland. Other wetland areas are smaller and, under U.S. federal definitions, standing water need be present for only a few weeks each year for an area to be protected as a wetland. Therefore, some wetlands are dry during the greater part of the year. When water is standing or the soil is saturated, wetlands are probably releasing methane into the atmosphere.

Production of methane by beef cattle has received much publicity, because cattle do generate a great deal of methane. Technically, anaerobic bacteria in the cattle's stomachs digest the grass eaten by cattle, producing methane. Many jokes imply that all methane is created when cows pass gas. The actual process is more complicated. Much of the methane comes from manure piles and droppings on the ground, but droppings of pigs, birds, or just about any animal produce methane. Cows also release considerable methane through belching.

Cows get the bad press for producing methane, but all members of the family of ruminants—animals that chew cuds—produce methane. Several domestic and wild animals are ruminants. Other animals that are not ruminants, such as pigs, also make methane. Some energy-efficient farms obtain methane for heating by collecting pig manure in a tank and using the methane that outgases from the manure.

Experts disagree about how much of the increase in atmospheric methane can be blamed on domestic livestock. Cattle raising has increased methane production, but that increase is partially compensated for by the numerical decline of other species of animals, such as elephants and buffalos.

Another major source of methane sounds like a joke. The source is termites. Cows convert eight percent of the carbon they digest into methane. Termites convert only one percent of the wood they eat into methane, but the termite population is much larger than the cow population.

Patrick Zimmerman of the National Center for Atmospheric Research issued a report on methane production by termites. He and his co-authors estimated that termites release 150 million tons of methane into the atmosphere each year. If true, this means that termites are the world's largest source of methane. A scientific reshuffling of the estimate then took place, as several researchers disputed Zimmerman's figure.

The disagreement had nothing to do with how much methane a single termite produces. It had to do with how many termites there are in the world. Throughout the argument, no one counted or proposed counting the world's termite population, although a German group measured the production of methane by individual termite mounds in Transvaal, South Africa. Even this group failed to count the number of termites that inhabited the mounds. However, the Germans changed the focus of the argument, which soon revolved around the number of termite mounds in the world. Someone finally suggested that termites produce 40 million tons of methane a year. When no one objected too loudly, the research community accepted this as an estimate.

Estimates often come about in this manner. Various experts guess, then negotiation takes place through scientific articles and letters to the editor about opponents' articles. Eventually, some value from the middle range of the guesses becomes accepted as truth. In many cases, no one attempts a measurement.

In this instance, there was so much interest in the methane-producing ability of the termites that the little bug's capacity for producing carbon dioxide was glossed over. Apparently, while termites convert just one percent of the wood they eat into methane, they convert 90 percent of that wood into carbon dioxide. According to Zimmerman, termites release 55 billion tons of carbon dioxide into the air each year, making them a respectably large source of carbon dioxide. If we reduce this estimate by a factor of about four—the same reduction factor everyone finally accepted for methane production—we conclude that termites still produce 14 billion tons of carbon dioxide per year.

As strange as it may sound, burning also releases methane into the air. Since methane burns so easily, the burning of biomass seems an unlikely source of methane. If burning were entirely efficient, all the methane would be consumed. But much burning, especially in agricultural fields, resembles smoldering rather than active burning. Heat causes some complex biomass molecules to break down as far as methane before they burn. Some of that methane escapes into the atmosphere because it hasn't gotten hot enough to burn, or because there was not enough oxygen present to induce burning.

ENVIRONMENTALISTS AND METHANE

Methane raises tough environmental questions. The fact that so much methane comes from the most natural of sources makes opposing its production sound somehow anti-environmental.

Rice farming may not be as natural as termites, but opposing rice farming does not have the same ring as opposing nuclear power. Opposing beef—especially growing cattle on federal grazing lands—seems fair sport, but even this position presents problems. Deer and elk produce methane, too. Should we also ban them from federal lands? How about the buffalo and the elephant? The herds of buffalo are gone right now, but they may be on their way back.

Commercial production of buffalos for meat is a growing business. Ted Turner, the owner of the television network CNN, owns a ranch that grows buffalo, and commercial operations may not be the only problem. There is a current movement that would like to see a large fraction of the Great Plains converted back into free range, so the buffalo could roam once again. Since herds of buffalo produce as much methane as herds of cows, it can be awkward to support such a program. But it can be an equally awkward program to oppose.

Wetlands also present contradictions. Wetlands harbor many kinds of life, but there is no denying that they also generate large amounts of methane.

Landfills provide one more problem for environmentalists. In fact, they may have been co-opted without realizing what has happened. The environmental movement advocating biodegradable trash ran into difficulties. First, it was demonstrated that a newspaper could still be read after it had spent 30 years in a dump. Then it was revealed that biodegradable trash produces methane. Even if some way existed to convert the methane into carbon dioxide, biodegradables would present a problem.

The buildup in carbon dioxide comes from burning fossil fuels. These fuels are carbon compounds that nature has stored up over millions of years. Burning them as fuel releases their carbon into the atmosphere, and these releases of stored carbon are what global warming is all about.

Compare this process to the destiny of trash. If the trash is biodegradable, it decays into either carbon dioxide or methane, both of which are greenhouse gases. On the other hand, when someone buries non-biodegradable trash, it lies inertly in the ground forever. While just lying there, it behaves like coal or petroleum. It is carbon in storage. Thus, biodegradable trash may contribute to the atmosphere's problems rather than resolving those problems.

An abundance of land exists for landfills, but there are two problems: the land is expensive; and trash must be hauled long distances in order to be dumped in available landfills. Cities are among the strongest advocates of biodegradable trash, because they might be able to fill and later reuse their expensive landfill sites after the trash decomposes.

Most scientists are sympathetic to the environmental movement. The scientists may raise environmental issues partly so they can get research funding, but they would also like a clean, natural world in which to live. For both reasons, they are often reluctant to attack or disagree with environmental activists, even when they realize that the activists are unknowingly presenting simple-minded solutions to complex problems. Nature's interactions are highly sophisticated. Even deciding whether plastic or paper bags are more friendly to the environment becomes a complex problem. The answer to most such questions is usually, "It depends."

IS AIR POLLUTION THE SOLUTION TO GLOBAL WARMING?

Air pollution is almost universally regarded as an evil—and rightly so. It contributes ugliness to our surroundings and leads to health problems.

On the ironically positive side, however, it can also counteract some global warming, and it offers a partial compensation for the ozone hole. Polluted air develops a significant ozone content. This ozone blocks part of the ultraviolet radiation that gets through the upper atmosphere's ozone hole. However, pollution is obviously not the solution to our problems.

Aerosols are small solid or liquid particles suspended in the air. They are too small to settle quickly to the ground, so they remain suspended for days or weeks. Some aerosols enter the atmosphere as dust or smoke particles; others are born when atmospheric gases condense and form particles. As table 2 shows, more aerosols come from natural sources than from human activities. Plants and dust whipped up by the wind are major natural sources. Pollutant gases interact and form other aerosols in the air. Sulfur dioxide produced when utilities burn high-sulfur coal is one example of a gas that quickly condenses and forms aerosol particles.

Most polluting gases are themselves colorless. The dirty color and haze associated with polluted air come largely from the aerosols it contains. The aerosols formed by condensing gases are smaller than the wavelength of infrared radiation and cause almost no greenhouse warming on their own. They can, though, contribute to a cooler climate by reducing solar heating. Less sunlight reaches the ground because the aerosols scatter light. Some is scattered back into space instead of being absorbed at the surface. Scattering light back into space decreases the amount of sunlight available for heating the ground.

	ESTIMATED PRODUCTION RATE (millions of tons/day)	TOTAL ATMOSPHERIC CONTENT (millions of tons)	PERCENTAGE OF TOTAL CONTENT
Aerosol Sources from Nature			
Dust rise by wind	1	16	24.1
Sea spray	3	7.6	11.9
Volcanic dust	0.01	1.6	0.2
Forest fires (intermittent)	0.4	6.2	9.9
Vegetation	3	17	25.8
Sulfur cycle	1	5.5	8.6
Nitrogen cycle ammonia	0.7	3.9	6.0
$NO_x \rightarrow NO_3$	1	5.5	7.7

TABLE 2 *Nature produces many more aerosols than men. Some aerosols remain in the atmosphere longer than others, so the total content and production rates are in different proportions.*

Clouds scatter light, too, but clouds and aerosols have different effects. The cooling effects of aerosols are simpler than the warming and cooling effect of clouds. Both scatter sunlight, but clouds block infrared radiation quite well, while small aerosols have little effect on infrared radiation. This allows them to reduce incoming solar radiation (ultraviolet) with little effect on the outgoing infrared.

Even aerosols can sometimes cause heating. When the ground surface highly reflects sunlight, instead of absorbing it, aerosols can increase greenhouse warming. Snow-covered areas and light-colored sands are examples of highly reflective surface cover.

Perhaps the most important effect of aerosols comes from the way they alter clouds. They serve as nucleation centers for cloud droplets, increasing cloudiness. That is, water condenses on the particles even before the humidity reaches 100 percent. This allows clouds to form where they normally would not.

Some aerosols make clouds appear whiter from above, and whiter clouds scatter a larger fraction of the sunlight back into space. Some clouds may look black from the ground, but if observed from above in an airplane, they all appear white. The aerosols that make whiter clouds apparently result from industrial pollution—low clouds over the Atlantic become lighter in the eastern Atlantic region, closer to Europe. Similar effects are in evidence over the Asian Pacific, near Japan.

Sulfate aerosols are the major cause of whiter clouds. Carbon particles, such as those seen in the exhaust from diesel trucks, have the opposite effect. These are quite black, and they continue to absorb radiation even after becoming embedded in water droplets.

Sulfate particles dissolve in water droplets, which prevents them from absorbing sunlight. That sounds desirable, but it produces a negative result. A sulfate particle dissolved in water forms sulfuric acid, and a dissolved nitrate particle forms nitric acid. After the droplet grows, it falls to the earth as acid rain. Some sulfates are natural, such as fumes from volcanoes, but industrial processes contribute most sulfates. Burning high-sulfur coal and smelting sulfate metal ores spews out tremendously large quantities of sulfates, which become aerosols. Australian researchers traced sulfates from the Broken Hill smelting operation for several hundred miles downwind by following the aerosol plume it formed.

Since some sulfates are created by burning coal and produce acid rain, it seems reasonable to pass pollution control measures that require reduced sulfate emission. However, even this can lead to one of those strange, counterproductive feedbacks that plague efforts to control pollution. Reducing sulfates would decrease the small sulfate aerosols that scatter sunlight back into space. The reduction of sulfates would also cause low-level clouds to become less white, and to scatter less sunlight back into space. Both of these effects would result in greater short-term global warming. Such an increase, though, would be a one-time event, instead of the continuously increasing greenhouse gases that come from burning fossil fuels.

Some pollutant gases absorb infrared wavelengths, which qualifies them as greenhouse gases. Ozone falls into this category. Other gases are not truly greenhouse gases, but their reactions can increase or decrease the concentration of greenhouse gases. They might be called "friends of the greenhouse gases," because they interact with gases that do cause global warming.

Air pollution chemistry is complex—too complex to cover here. However, a short explanation of just a few processes can illustrate important effects. For example, the hydroxyl radical, OH, interacts with many gases—destroying some, helping others form. Interaction with OH is the primary destruction mechanism for methane. The same OH controls ozone formation through its interaction with nitrogen compounds. These are not the only interactions of ozone and OH, but they begin to indicate the complexities

involved. The result is this: Introducing one new pollutant can affect the concentrations of other pollutants. A new pollutant introduces new reactions that may speed or block other reactions. The net mix of gases in the air is altered.

NITROUS OXIDE

Nitrous oxide and methane absorb radiation at the same infrared wavelengths. Both block radiation from escaping through the atmosphere. In a pound-by-pound comparison, nitrous oxide produces 300 times more greenhouse heating than carbon dioxide. However, the small quantity of nitrous oxide in the atmosphere means the total heating it generates is only four percent of what is produced by carbon dioxide.

Nitrous oxide originates largely through natural processes, much like methane, but the processes are different. The atmosphere currently contains about 310 nitrous oxide molecules in each billion molecules of air. That may not seem like much, but it represents an increase from 285 parts per billion in pre-industrial air. Samples taken from ice cores show that the increase started around 1700. The current increase is 0.25 percent per year, considerably less than carbon dioxide's 0.5 percent and methane's 0.9 percent-per-year rate of increase.

Nitrous oxide is most commonly used as a dental pain killer. It was once popular as a recreational gas, given the nickname "laughing gas." Parties were devoted to the hilarity that resulted from breathing nitrous oxide. According to one story, a dentist saw a man injure himself while high on nitrous oxide. The man felt no pain, which gave the dentist the idea of having his patients breathe the gas in order to avoid pain. If nitrous oxide in the atmosphere increases sufficiently, the world might die a happier but hotter place.

As Table 3 indicates, there is about a 50 percent uncertainty in our estimates of the amount of nitrous oxide that comes from different sources. As is the case with methane, much of our ignorance is due to the fact that nitrous oxide is produced naturally. The production budgets of industrial chemicals and their by-products are relatively easy to estimate, because economic reports include production figures. But natural production rates of most chemicals are more difficult to judge. Production usually occurs over large areas, with only trace amounts injected into the air from each square meter of land or sea. Production rates of chemicals that come from the soil, like nitrous oxide, vary with soil type, temperature, and rainfall. Accurate estimation of production requires taking measurements in varied locations under all weather conditions. Arriving at such a set of measurements throughout the world would be terrifically expensive.

The ocean produces a third of the nitrous oxide that gets into the atmosphere. Questions about which part of the ocean the gas comes from remain unanswered. An increase of nitrogen in surface waters may be producing nitrous oxide, or the nitrogen could be coming from deep in the oceans, where little oxygen exists.

Up-welling waters seem more saturated than surface waters, which suggests production occurs in the depths.

Soils are the next-largest source. The soils in tropical forests generate more nitrous oxide than those in temperate forests. Soils where tropical forests have been cut sometimes triple their release rate. To quote the IPCC: "Reliable global N_2O [nitrous oxide] fluxes from grasslands are impossible to derive from the fragmented data available."

Many people have tried to correlate nitrous oxide production with the application of chemical fertilizers to soil. Because many fertilizers are nitrogen compounds, the connection seems obvious. But estimates disagree dramatically. Those considered credible by the IPCC indicate that somewhere between 0.2 and 50 percent of the annual world increase of nitrous oxide is attributable to fertilizers. In other words, fertilizers may be responsible for almost none or about one-half of all excess nitrous oxide! This scientific disagreement undoubtedly needs to age for a while before everyone settles on a mid-range number.

A few years ago, experts thought burning of forests, agricultural wastes, and

	RANGE (Millions of tons/year)
Nitrous Oxide Sources	
Oceans	1.4 – 2.6
Soils (tropical forests)	2.2 – 3.7
Soils (temperate forests)	0.7 – 1.5
Combustion	0.1 – 0.3
Biomass burning	0.02 – 0.2
Fertilizer (including groundwater)	0.01 – 2.2
TOTAL	4.4 – 10.5

TABLE 3 *Oceans and soils probably produce most of the nitrous oxide in the atmosphere, and chemical fertilizers could be another major source. The amount absorbed by soils is unknown, but we know the atmospheric concentration is increasing.*

fossil fuels was the largest source of nitrous oxide. Until ten years ago, estimates of nitrous oxide produced in this manner were 10 to 100 times too large. Such huge errors occurred because researchers used glass flasks to collect air samples near fires, then carried the flasks back to the laboratory for analysis. Until the 1980s, scientists failed to follow the number one rule of experimental measurement: always do the null experiment—analyze a control sample and see if it gives the right answer. After someone finally performed this important step, we realized that nearly all the nitrous oxide collected in flasks had been generated within the flasks themselves. All previous measurements suddenly became meaningless, and burning was reclassified as a minor source of nitrous oxide.

The stratosphere is the sink for nitrous oxide. After it is produced at the surface, the gas eventually rises through the tropopause into the upper atmosphere. Again, this is similar to what happens to methane. Intense solar ultraviolet radiation in the stratosphere breaks nitrous oxide up, just as it breaks up CFCs. On the average, a molecule of nitrous oxide lasts 150 years before it reaches the stratosphere and is destroyed.

The IPCC report contained an interesting conclusion about the increase in nitrous oxide. The panel admitted that it was "difficult to account for the annual increase based on known sources." Still, the IPCC said it believed that human activities caused the increase, and called for a "reduction of 70 to 80 percent of the additional flux of N_2O." In effect, the panel said, "Whoever is generating the stuff better stop."

Nitrous oxide is only one of several nitrogen-oxygen compounds. This whole family of compounds is often called NO_x. The "x" suggests any member of the family of molecules that contains only nitrogen and oxygen molecules. Members of the

family other than nitrous oxide exist for only a short time in the air, because they quickly react with other molecules.

OZONE AS A GREENHOUSE GAS

Ozone serves as a shield that protects us from the sun's intense ultraviolet, but it also absorbs solar energy in the visible spectrum and thermal radiation from the surface at infrared wavelengths. The concentration of ozone in the air varies with altitude, and from one period to the next. All this activity makes ozone a very important player in the atmosphere. It also means that ozone affects temperatures in circuitous ways.

In the stratosphere, at altitudes above 60,000 feet, we've seen that temperatures increase with altitude. The air gets warmer because ozone absorbs solar ultraviolet energy. Heating is greater at the higher altitudes where the ultraviolet is more intense, and is not yet depleted by the ozone's absorption. That same ultraviolet contributes to the production of ozone, so the higher altitudes contain a larger fraction of ozone for absorbing the solar radiation.

By the time the solar beam gets to the bottom of the stratosphere, the ultraviolet radiation is too weak to provide any more heating. Warm air from the surface cannot rise to such high altitudes, so temperatures remain unchanged over some range of altitudes. This constant temperature region, you'll remember, is called the tropopause.

Even after ozone absorbs all the ultraviolet, it continues to absorb radiation in the visible part of the spectrum. The absorption never depletes the radiation in any wavelength region, but it does increase air temperatures. Water vapor and carbon dioxide also absorb a small amount of solar energy and heat the air at wavelengths too short to affect radiation losses from ground emissions.

Ozone plays a double role in the lower

atmosphere. It warms the air by absorbing incoming solar radiation, and it stops infrared cooling by blocking surface infrared emissions. Warming the air, strange as it sounds, reduces greenhouse heating because warm air emits more infrared radiation of its own into space, thus ridding the world of heat. Blocking the infrared emissions from the surface has the opposite effect. It causes greenhouse heating. The ground recaptures some energy that would otherwise go back into space.

Air pollution usually gets blamed for ozone in the lower atmosphere. In fact, authorities often base the index of air pollution on ozone content. But urban air pollution is not the only source of hydrocarbons leading to ozone pollution. It is unfair to blame automobiles and industrial processes for all the hydrocarbons. Trees and plants release large quantities as well. The blue haze in mountain air is due to molecules from the alpine forests. When conditions are just right, very large concentrations of surface level ozone can occur in the countryside, far from urban sources. Ozone levels in the countryside sometimes get as high as those seen in polluted urban air. While city officials use ozone levels as the criterion for declaring air pollution crises, few people realize that on this basis, the pristine country air sometimes qualifies as too polluted for children to play in. Such conditions are unusual, but they do occur.

The amount of ozone formed at low altitudes depends on the number of hydrocarbons and the amount of nitrogen-oxygen compounds present in the air. Hydrocarbons come from evaporated gasoline, poor combustion in automobiles, and burning. Natural sources, such as growing plants, also exist. Polluted air has more hydrocarbons, largely because of automobiles. The presence of excess hydrocarbons explains why ozone increases as the air becomes more polluted.

Even as the CFCs destroy ozone in the upper atmosphere, ozone is increasing in the lower atmosphere. Some increases come from greater amounts of ultraviolet that reach the lower atmosphere, and some from increases in methane, other hydrocarbons, and nitrogen-oxygen compounds. Researchers think the increase may be about one percent per year. Therefore, surface concentrations in Europe may be double what they were only 50 years ago. But ozone concentrations vary so much with location and time frame that the real rate of increase is highly uncertain. Low-altitude ozone is destroyed within days through contact with plants or chemical interactions.

The ozone in the lower atmosphere serves as a greenhouse gas. Because it is concentrated near the surface, though, it has low effectiveness. The fact that it warms the air by absorbing short wavelengths decreases its effectiveness even further. The net result is that low-level ozone is not a very significant greenhouse gas.

Near and above the tropopause, ozone becomes a major greenhouse gas. We need the ozone to block the sun's ultraviolet, but that same ozone causes considerable greenhouse heating. This heating is seldom discussed, because natural processes (unrelated to humans) create the ozone. However, humans are now destroying some of that ozone, and decreasing the high altitude ozone also decreases global warming.

Here we meet paradox once again. We finally find a greenhouse gas whose level is actually decreasing, but the resultant loss of ultraviolet protection leaves us vulnerable.

8

The Worst-Case Increase in Temperature

Remember, you gave up your air conditioner to save the ozone.

News releases about global warming often have a cataclysmic ring. There is an underlying "end-of-civilization" message that can induce a touch of terror. The reports never predict what life will actually be like in 100 years; they simply offer varying predictions of change—and change is fearsome for most of us.

Rather than reacting to wild and inconsistent speculations, perhaps we should look the beast right in the eye, examine the worst-case scenario, then decide whether the end of civilization is just around the corner. Change can be unpleasant, but if we can handle the worst case, there is no longer any need to fear the unthinkable unknown. And whatever really happens will undoubtedly be more desirable than the worst case.

If the concentration of carbon dioxide doubles, the consensus among experts indicates we can expect temperature increases of from 3 to 9°F (2 to 5°C). At the present rate of increase, that doubling of carbon dioxide and the corresponding temperature increases would happen over about 100 years. Thus, we might assume that the worst threat involves temperatures nine degrees hotter than current levels. What are the implications of such a change?

In the United States, most of us are isolated from our climate. We have air conditioning as well as heating, often in our automobiles as well as our homes. Our primary weather worry has more to do with the effects of snow and rain on traffic than with becoming personally wet or cold. In unusually bad situations, we may have to follow some schedule that tells us when we can water the lawn, or we might be asked to adjust the air conditioner in order to prevent an electrical brown-out.

We are not immune to the weather, though. It can affect us in significant secondary ways. Climate changes could threaten our food supplies. An increase of nine degrees in temperature is not enough to stop all crops from growing, but it might force farmers to switch to crops that produce well in warmer climates. Changes in rainfall are actually more important than increases in temperature. If rainfall holds constant as the weather warms up, growing seasons will be longer, and agricultural production might well increase. When rainfall decreases, agricultural production drops.

Famine currently threatens some areas of the world, but on an overall basis, the world actually has a surplus of food. The United States and European Community nations have been engaged in a trade war over agricultural exports for a number of years. Each year, the U.S. spends about one billion dollars in subsidizing agricultural exports. These programs are in place to prevent Europe from taking over markets, leaving us with no place to sell our surplus farm production. Canada, Australia, Brazil, and Argentina also produce large agricultural surpluses.

Those of us fortunate enough to live in the industrialized world have little to fear from famine. Crops may change as the weather changes, but our economies are agile and capable of rapid change. Less industrialized societies susceptible to famine may find change difficult or impossible. A nomadic, pastoral society may require generations to adapt when a major change is inflicted upon its world. In American society, every generation essentially grows up in a different culture. Each generation uses products that did not exist in the previous generation; each generation even eats different foods.

As we've seen, the most pessimistic scenarios indicate that it should take 100 years for temperature to increase by nine degrees. A hundred years is hard to imagine, because we tend to measure time by changes in our bodies and our jobs.

It helps to think back 100 years. In 1889, people in covered wagons lined up at the Oklahoma border for the first big run to claim homesteads. Less than 50 years later, the Dust Bowl famine drove half of those settlers off their farms. The change in climate occurred just as California started benefiting from the irrigation projects that transformed it into our largest producer of fruits and vegetables. Many "Okies" adapted to the change of climate by heading to California. Such migrations have occurred throughout our history. Economic expansion has made us a dynamic society, and climate changes may force us to continue in that mode.

All of the events above took place in less than 100 years. Agriculture can and does change. Farmers must continually adjust what they grow as the climate varies and the market for some products declines. Cotton was king in the South for only about 100 years. South Carolina no longer depends on indigo as a major cash crop, and who would even recognize the field of flax once needed for the weaving of linen? If the amount of rain decreases over the next century, some regions may have to adjust, but they are capable of doing so.

Land use in Oklahoma is certainly different today from what it was just before the Dust Bowl. Then, farmers tilled the land with row crops, because homesteaders had plowed up the prairie sod. The Dust Bowl caused land use to change. Much of the land has come full cycle, and once again grows grass. Now cattle graze instead of buffalo. This cycle took place in just 100 years, and it serves as an example of the degree of change we can accommodate over the course of a century.

There is another often-stated concern related to rising temperatures—that the polar ice caps will melt and sea levels will increase. Increases in sea level may not worry people living in Denver, but people living on a South Pacific atoll have more at stake. Their whole island might disappear.

In considering the worst case, increases in sea levels seem to offer the greatest potential disruption. Experts now expect increases in sea level of no more than three feet. In the last 100 years, sea levels have increased nine inches—fully one-quarter of the worst-case expectations. The effects of the increases have been so small that most people are unaware that any increase has occurred.

Predicted increases in sea level would submerge a small fraction of the world's land area. Loss of the land itself would not be too important. Cities are the main problem—the land lost would include parts of many major world cities. To counter the problem, governments would no doubt be forced to build dikes around cities to hold the water out. Changes in rainfall could produce major economic dislocations, and the construction of dikes is an expensive undertaking.

Perhaps the greatest hazards we face are the economic disruptions that can come from changes in climate—think of the Great Depression of the 1930s. But, on the other hand, 100 years is a long time. Changes spread over such a long time might be less painful than abrupt economic change. No one claims that major economic disruptions would be fun, but they would be survivable.

We must have confidence that we can survive the changes ahead. The fact that we can stand whatever might come, of course, does not mean we shouldn't take action. Even if our actions fail, we will still be around; in the meantime, we should continue to search for ways to prevent changes that will cause pain and suffering. Coping with climate change would cost billions of dollars more than changing our ways now and reducing emissions of greenhouse gases.

CLIMATE EQUILIBRIUM AND TRIP POINTS

Equilibrium occurs when all the forces acting on an object cancel each other out, leaving the object balanced. Equilibrium does not mean that there are no forces acting on the object. It is similar to a tug-of-war—both sides pull, but neither side moves. If one side pulls a bit harder than the other, the lack of balance destroys equilibrium, and movement occurs.

The climate never quite reaches equilibrium. If it did, the weather would repeat itself year after year without change. The climate is like a tug-of-war with well-matched but not perfectly balanced teams. Many forces act to move the weather in different directions, but none prevail. Change begets change, and the climate never reaches equilibrium. Still, the climate changes slowly enough so that it is never very far from equilibrium.

We call the forces that act upon our climate "feedbacks." When change occurs, positive feedbacks tend to make the change larger; negative feedbacks make the change smaller. Negative feedbacks tend to resist change, to force the weather back, to counteract change. They give the climate stability. Positive feedbacks can lead to oscillations—periodic repetitions such as El Nino or droughts. The atmosphere is full of such oscillations, and they may interact in complementary or counteracting ways. Sometimes the interfering cycles reinforce each other and become more intense. At other times, they tug in opposite directions and make cycles like drought or El Nino almost unnoticeable.

Water provides a good example of feedback mechanisms. When the weather heats up, more water evaporates from plants. Water is the strongest of the greenhouse gases, so the additional water vapor traps more greenhouse radiation, which causes the air to warm further. The increased water vapor provides a positive feedback. Heating caused more water to evaporate, and the additional water vapor caused more heating.

Clouds can act as a negative feedback. The additional water vapor that evaporates when the air warms can rise to higher altitudes, where the air is cooler. Cooling can cause the water vapor to condense and form a cloud. Clouds block the sun and let the surface cool. Thus, heating produced increased evaporation and led to clouds, which allowed the surface to cool. The clouds counteracted the heating, creating a negative feedback.

If carbon dioxide increases and nothing else changes to compensate, everyone agrees that greenhouse heating will take place. The trouble is, other things will also change. The argument becomes one of positive and negative feedbacks. Those expecting large accompanying changes point to positive feedback mechanisms and say, "Those feedbacks amplify greenhouse heating." Climatologists are aware of several positive feedback mechanisms. On the other hand, those who doubt climatic change can suggest only a few negative feedback mechanisms. This may just mean that no one to date has recognized all of the negative feedbacks. The biggest feedback of all—loss of heat through radiation—is negative. When the weather warms, radiative losses increase. The greenhouse effect has everything to do with the way in which carbon dioxide and other gases cut down the effectiveness of this negative feedback mechanism.

Equilibrium can be stable or unstable. A marble in a trough of corrugated material is in stable equilibrium. Bump it away from the lowest part of the corrugation, and it oscillates back and forth, finally coming to rest at its starting point. This is an example of how negative feedback can produce stability. Balance the marble on

the high ridge between the troughs, and it is in unstable equilibrium. A small push to either side brings positive feedback into play. Once disturbed, nothing pushes back to reestablish the marble's balance at the ridge's high point. The marble rolls down the side of the corrugation, oscillates a bit, then comes to rest at the bottom of a trough. Once disturbed from an unstable equilibrium, the marble finds a different equilibrium point.

Climate changes so slowly that changes never force climate far from its stable equilibrium point. Temperatures do not vary radically from the norm, although slow increases or decreases take place. The atmosphere does apparently have some unstable regions, though.

Ice ages may well be unstable points. Ice ages come along, but these periods are shorter than the warm, stable periods in between. The low temperatures during ice ages seem to trip some mechanism, ensuring that the world warms up again—a warming that takes thousands of years. Some other kind of trip mechanism apparently exists as well, one that eventually upsets the equilibrium and drives the earth into another ice age. Ice ages are one of the climatic cycles produced by positive feedback.

Some feedbacks have a trip point, meaning they do nothing until a certain point, then become active. Dew is one example. The night cools until the air becomes saturated with water vapor. At the dew point, water begins to condense on plant leaves and releases heat. Condensation is the trip point. The heat released by the condensation of water vapor provides negative feedback that slows the cooling rate. Trip points that provide negative feedback seem desirable, in that they provide stability. Trip points that provide positive feedback seem undesirable, in that they speed the onset of changes like ice ages.

At least one trip point exists that can lead to catastrophic changes through positive feedback. In the permafrost of Alaska and Siberia, there are the large quantities of methane condensed into solids called methane clathrates, which look like wet snow or ice. These methane clathrates contain about six times as many water molecules as methane molecules, but they still hold a great deal of methane. According to estimates, the permafrost of Alaska and Siberia stores 400 billion tons of methane.

If the temperature should warm several degrees, the frozen clathrates might vaporize into the atmosphere. This would spell real trouble, because methane is many times more effective as a greenhouse gas than carbon dioxide. The evaporation of large quantities of methane could greatly increase greenhouse heating, thus causing even higher temperatures. Higher temperatures, in turn, would cause more methane to evaporate. According to this scenario, a little warming could drive the atmosphere into runaway temperature increases. Great uncertainty exists about how much warming the permafrost can endure before the clathrates start releasing methane.

The polar caps can serve as another example of the trip point mechanism. If the polar regions become warmer, no melting can occur until temperatures rise above the 32°F trip point—the temperature of melting ice. Present estimates indicate that temperatures will not increase to that degree. However, if temperatures exceed the trip point, sea levels might rise much more than predicted. If the trip point is not reached, global warming might even cause an increase in the depth of the polar ice pack, because warmer air holds more moisture and can produce greater snowfall.

Trip points are difficult for scientists to deal with when they are trying to estimate how hot the earth might become. Trip points may exist that we have not yet recognized, because the conditions necessary to trigger them have never been satisfied. For example, if methane were not of commercial interest because it is used as a natural gas, we might not be aware of frozen methane in the permafrost. In that case, we would certainly be in for a big surprise if heating warmed the permafrost and atmospheric methane suddenly doubled or quadrupled.

Ocean circulation patterns must not be overlooked when discussing trip points. The underwater circulation patterns are virtually unknown—and we know even less about the forces that determine the patterns. Small shifts in ocean currents could produce rapid shifts in climate, as exemplified by El Nino. Under this recurring weather pattern, a small temperature shift in the Pacific, off of South America, produces drought in the corn belt, floods in Texas, and significantly different summer temperatures throughout the midwest. If the Japanese current shifted, California would lose its extended growing season. It could no longer provide a large fraction of the nation's fruits and vegetables. Shifts in climates normally take place over many years, but ocean currents could shift in weeks or months and transform climates almost as quickly.

In discussing what scientists do and do not know about the atmosphere, we must acknowledge the tendency of scientists to exaggerate—not as individuals, but as a group. They exaggerate primarily by failing to explain to the public how little is known and how poorly they can predict what may happen to the earth. Do not, however, take this to mean that the predictions of global heating should be ignored. When it comes to details, the predictions are probably incorrect; they may even be wrong in certain gross aspects; but we must not ignore them. They are the best available predictions based on current knowledge, and as such they provide the best possible guess about what we can expect in the future. It is important to bear in mind the warning of Steven Schneider, a climatologist at the National Center for Atmospheric Research. As Schneider says, "The models are as likely to underestimate as to overestimate the warming."

Trip points are the most unpredictable variable in contemplating climate changes. If we allow just a little climatological change, we may trigger a trip point that could cause runaway heating. And trip points are not limited to temperatures. They could affect rainfall, winds, or ocean circulations. Since no one knows all the trip points, permitting any degree of climatic change amounts to playing Russian roulette. In all probability, nothing will happen; but if the improbable happens, the results could be devastating.

WHAT THE COMPUTER MODELS SAY

Ask a meteorologist how much warming to expect, and the answer may be none, some, or a lot. A particularly loquacious weatherman might speculate about the geographic effects in particular regions. Generalities prevail because meteorologists know that climatic processes are too complex to allow for specific predictions. They realize there is simply no chance that a single equation can specify the answer, or that someone can work the future out on a calculator. A computer model, though, behaves differently. In essence, the model does not recognize its own limitations. It spews forth highly detailed answers without worrying about whether the projections might be wrong.

Global change projections must nevertheless come from a computer model, because the processes are too complicated for any other source. Since this is the case,

it seems feasible that the bigger and more complex the model, the greater the chance that it will offer correct information. Several of the larger current models are of comparable size and complexity. A consensus among these models might provide reassurance that their projections are correct.

The International Panel on Climate Change (IPCC) conducted an experiment. It ran several different models under identical assumptions and compared the results. All the models were asked to predict what would happen if carbon dioxide concentrations suddenly doubled. Since the models were instructed to allow for no clouds in their calculations, all the models' answers generally agreed. Models that don't allow for clouds may not seem realistic, but the discovery of a reasonable degree of agreement between models elated the modelers. The agreement disappeared, however, when the first set of clouds was added. Some general qualitative agreement existed, but temperature and rainfall differed widely from model to model.

The best consensus possible on model results indicates that a doubling of carbon dioxide will cause surface temperatures to increase from 3 to 9°F. Speakers and the media often quote these numbers, but they are almost meaningless since they do not indicate where or when we can expect the heating. The models actually agreed more strongly regarding where and when the hotspots would occur than on how hot they would become. The good news is that most of the heating will occur in the winter. Even better, the heating should be greater in the northern latitudes and Canada than in the middle latitudes.

If someone had suggested a weather modification scheme to accomplish just this kind of climate change, world governments might well have spent billions to implement it. Warmer winters would extend the length of the growing season in the northern states and Canada, making new and higher-yielding crops possible. In the southern states, the warming would be a little more than half as great, meaning production could double in areas where warmer weather made seasons long enough to grow two crops a year. The mitigating bad news is that the models predict decreased rainfall over most of the United States. Less rainfall means reduced agricultural production, because farmers usually grow crops that need a little more than normal rainfall. If global warming decreases rainfall, current crops will suffer from lack of water. Farmers will be forced to shift to crops that require less rain. Worldwide, however, the models predict increased rainfall. This indicates generally increased agricultural production in most of the world.

The models also predict changes in the seasonal distribution of rainfall. Warmer air's ability to hold more water may explain why the models predict more rain in winter. Less rainfall and hotter weather in the summer could decrease soil moisture, a development also predicted by the models. However, this prediction has credibility problems. The models did not consider changes in vegetative cover or land use, even though loss of soil moisture is critically dependent on both factors.

The failure to address vegetative cover and land use is a major shortcoming of the model predictions, even those predictions not directly related to soil moisture. All models assumed an instantaneous doubling of carbon dioxide and no change in surface vegetation, which does not remotely resemble the way in which the world would change.

Vegetation and climate share a strong co-dependence. As the climate becomes drier, plant cover starts disappearing. Solar heating of newly bared ground accelerates drying and loss of plant cover. Decreasing plant cover in low rainfall areas provides a positive feedback, pushing those areas toward desertification. Plants affect temperature through evaporative control, and by shading the ground and keeping it cooler. The effect of plants on climate is so strong, in fact, that climate cannot be specified without knowledge of ground-level vegetation. As climates change, plant cover changes and vice versa. Rain forests themselves produce the conditions that cause rain forests. This, of course, means that cutting rain forests may stop the rain.

We are not currently able to create feedback between plant cover and climate within a computer model. The problem is twofold. First, such coupling would require too much computer time, even on the fastest computers. Second, computer modeling of plant growth and competition between plants is still an infant field. We have not developed enough skill to include plant growth in a realistic way, even if the computers were fast enough. Plant models are even more complex than weather models, so such models would need massive amounts of computational time. Ten to 20 years will probably pass before models can even approximate the interactions of plants and climate on a global scale.

In the past, as the world warmed and cooled, climate changes occurred over thousands of years. Plant populations were afforded sufficient time to adapt, so the plant-climate system was never far from equilibrium. But the plant community cannot adapt fast enough to keep up with the rapid temperature increases expected from greenhouse warming.

As an example, predicted temperature increases over the next 100 years are great enough to cause the north-south boundary between beech and pine forests to shift by at least 100 miles. Forest boundaries would thus have to move by a mile each year, but temperate zone forests simply cannot move that fast.

Until computer models incorporate plant competition, no one can reliably predict how plant cover might be affected by warming. Forests adapt to climate changes more slowly than they recover from a forest fire—and recovery from fire can take over 100 years. In the case of a forest fire, the driving event (the fire) is quickly completed, so recovery can start at once. In changing climatic conditions, the driving events may remain in effect for a long time, thereby delaying the recovery.

Atmospheric scientists treat models with suspicion. When it comes to estimating climate, however, the same scientists admit that models provide our best available predictions. As one scientist, Pat Squires, complained when he was working at the Desert Research Institute, "We can't afford to trust models, but we can't afford to ignore them, either."

HAS GREENHOUSE WARMING STARTED?

The debate about whether greenhouse heating has started certainly heats up scientific tempers. Some scientists claim that errors invalidate our historic temperature records. Others claim that some factor other than the greenhouse effect made the 1980s the warmest decade on record. The issue is almost as controversial as evolutionary theory was in Darwin's time, but it is actually of minimal importance.

The question of whether greenhouse gases have caused the over 100-year-long upward trend in temperatures is ultimately one of belief. As in Darwin's time, the intensity of some researchers' beliefs approaches a religion. Those who believe that the greenhouse effect is a clear and present danger find both validation and warning in the increasing temperatures.

We might consider them "believers." We can apply the label "non-believers" to those who dispute the idea that carbon dioxide increases contribute to present or future global warming. Earth's century-long warming trend puts non-believers on the defensive, so they generally discredit any suggestion that links the warming trend with greenhouse gases.

Whether greenhouse gases contribute has become an article of faith for both groups. If both sides were less evangelical, they might begin to debate the real question: How much warming can we expect, and when might it occur?

Non-believers operate under the unstated assumption that something keeps a lock on our climate. This lock *must* be an assumption because, as we noted earlier, virtually all experts acknowledge that the earth would be much colder without greenhouse heating. If we stripped the water vapor and carbon dioxide from the atmosphere, the earth would become a colder planet, more like Mars. If we increased the concentrations of methane and carbon dioxide, the earth would become more like Venus, which has a surface temperature hot enough to melt lead. Without natural greenhouse warming (not caused by man), the earth would be 60°F colder—so cold that life might never have evolved here.

We know the history of atmospheric temperatures all the way back to the ice ages. Disagreements exist about what caused temperature changes that led to ice ages, but no one doubts that the ice ages really occurred. Because history tells us that a wide range of temperatures has existed on earth, it seems to refute the idea that something "locks" temperatures in place. Perhaps those who deny greenhouse warming are taking an extremist stand in order to compensate for those on the other extreme, who insist that the future will be intolerable. As usual, the truth seems to

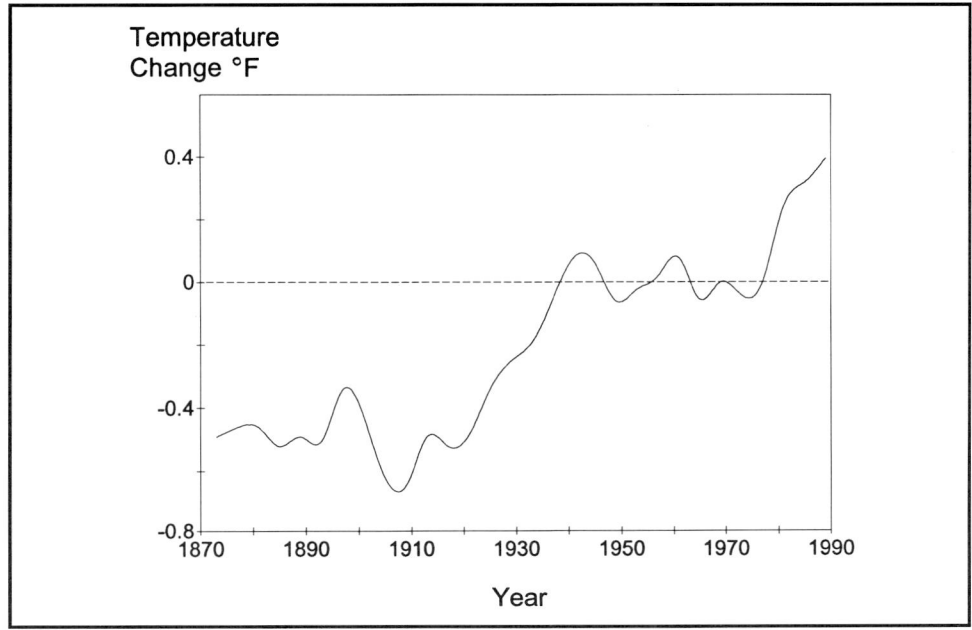

FIGURE 1 *This curve presents the main argument for the belief that global warming has started. Notice that temperatures have increased, but in steps rather than continuously. From 1870 until 1925, there was no net warming. From 1940 until 1975, temperatures actually decreased a little. Still, it is hotter now than it was in 1870.*

fall somewhere in between.

The temperature history shown in Figure 1 is at the center of the argument. If we look at the levels between 1860 and the present, it is difficult to claim that temperatures are not increasing. Further, comparing plots of temperatures and carbon dioxide levels over thousands of years makes it seem obvious that there is a link between the two.

Those on both sides of the argument offer good points, and ultimately there is no way to tell who is right. Remember, however, that the argument is a red herring that distracts from the real question. Probably the best answer we can come up with is this: Some greenhouse heating has occurred, but normal temperature fluctuations and measurement errors make it impossible to tell how much of the heating is due to human activities. We know that

Earth can heat up on its own, without human assistance. The difference today is this: We now have reason to suspect that humans are causing the warming.

THE NON-BELIEVERS' ARGUMENTS

The temperature fluctuation argument goes like this. Look carefully at the temperature levels in Figure 1 and you will see a temperature decrease from 1865 to 1895, and another decrease between 1940 and 1970. Increases in the 1980s were so rapid that experts believe greenhouse heating could not be the sole explanation. Until the 1980s, the temperature trend actually had experts wondering if another ice age was starting. According to the historic cycle, another ice age is due within the next few thousand years. Arguments against greenhouse heating ascribe the higher temperatures in the 1980s to a drought in the U.S. and the fact that El Ninos caused increased tropical temperatures.

The believers respond to the temperature fluctuation argument by acknowledging the existence of fluctuations. The fluctuations caused the two extended cooling periods. However, the data also indicates that a definite upward trend began with the onset of the Industrial Revolution, when we started injecting large amounts of carbon dioxide into the atmosphere. It seems highly unlikely that the correlation between carbon dioxide increases and global warming is a coincidence.

Non-believers also appeal to our nationalistic biases. We live in the United States, and our real interest lies in what happens to this country, not whether the world is heating up. Believers grudgingly admit that the United States has seen no warming trend during the last 100 years. But they insist that worldwide temperatures, not American temperatures, are the important measure.

Of course, the argument is more complex than this brief review. A close look at the data reveals that the western part of the United States has warmed slightly, while the eastern part has cooled. The net result is no change in the average temperature for the contiguous United States. Some even claim that the United States has cooled off, and that a phenomenon known as the "heat island effect" biases the temperature record.

The heat island effect is real. Cities act as "heat islands" because they contain so much black asphalt and so many black roofs. These absorb sunlight, making the cities hotter than grass- and plant-covered areas. Plants have leaves that provide cooling through evaporation. They absorb sunlight and convert the absorbed energy into stored chemical energy—not into heat. The cities also get hotter because

buildings block the flow of air that would normally ensure that hotter surface air is mixed with cooler air from above.

Those who claim that temperature records contain systematic errors point to urban areas and the heat island effect. They claim also that urban areas contain more than their share of temperature measuring stations, thus biasing the temperature records. They have a point—a city of as few as 10,000 people acts as a heat island. Greenhouse believers respond by saying that the heat island effect may exist, but its influence is small.

Another non-believer argument holds that there are too few temperature recording stations in many areas of the world, so we cannot arrive at proper calculations of worldwide average temperatures. With poor representation from large areas of Asia, Africa, and the oceans, who can say how world temperatures are really changing? Since the records indicate that the U.S. has experienced no net heating, the same might be true of other areas.

Some parts of the world are drying out and undergoing desertification. This is a slow process that is apparently unrelated to greenhouse heating. A study by Robert Balling at Arizona State University concluded that temperature recording stations in areas that are undergoing desertification see ever higher temperatures. Balling says this temperature increase is non-greenhouse warming, and he points to these stations as one reason that average world temperatures now appear to be rising. According to Balling, the world may be getting hotter, but we shouldn't blame it all on greenhouse gases.

Non-believers also cast doubt on the accuracy of older temperature measurements. This argument is a little strange, because errors tend to cause temperature readings that are too low as well as too high. Averaging the errors generally indicates that errors in both directions balance out, leaving us with with relatively small net errors.

A recent change in the method of measuring temperatures illustrates the difficulty we can nonetheless encounter. The U.S. National Weather Service has a network of over 5,000 cooperative observing stations. In the cooperative network, volunteers routinely record temperatures and other weather data. Between 1984 and 1990, the Weather Service introduced a new type of minimum-maximum thermometer at 60 percent of its volunteer network stations.

Scientists at the National Climatic Data Center analyzed the differences between minimum-maximum temperatures using the new and old instruments. Changing to the new instruments caused a 0.7°F decrease in the daily maximum temperature and a 0.5°F increase in the daily minimum temperature. Innacurate calibration of the thermometers was not responsible. Instead, the differences resulted from changes in the measurement locations. Changes in the way the thermometers were mounted might have added to the changes. It is hard to determine which method of measurement is better, or which is in error. The two methods are simply different, and so they arrived at different results. Still, the differences are good ammunition for those who claim that historical temperature measurements contain systematic errors.

Roy Spencer and John Christy of Huntsville, Alabama, confounded both sides of the argument in their report on satellite-based temperature measurements. These researchers obtained nearly complete coverage of the earth by comparing atmospheric sounding data obtained from TIROS-N satellites. The satellites scanned the entire earth each day between 1979 and 1990.

The satellites measured average temperatures over a large area and in the lower levels of the atmosphere. Given this measurement technique, direct comparison with results from ground stations is impossible. Ground stations measure temperature at a particular point, while satellites average temperatures over the area in their field of view. There is no way of determining how the two sets of measurements should compare. However, satellite averages and ground station averages should at least go up and down together. This would indicate that the averages obtained from the ground station network represent the world as a whole. Unfortunately, the two did not go up and down in quite the same way—indicating that the worldwide ground station network provides less than complete coverage. However, the coverage does not appear flawed enough to cause large errors.

The bombshell contained in Spencer and Christy's report was their conclusion that no temperature trend emerged from the ten years of data. The researchers admitted that temperatures were warmer toward the end of the 1980s, but they ascribed the difference to normal variability and the effects of El Nino. Their results really said little about whether greenhouse warming has started. We must also acknowledge that ten years is probably too short a period to reflect a true climatic trend.

These are only a few of the arguments that take place between the believers and non-believers. Others include claims that higher temperatures on Earth result from increases in the heat from the sun, or from changes in cloud cover. Most of these arguments are complex and unresolvable. For example, the non-believers point to the fact that temperature increases are greater in the southern hemisphere, and suggest that the southern hemisphere should not have experienced more warming, because most carbon dioxide releases occur in the northern hemisphere. It's an interesting observation, although no one knows how significant it is. Thomas M. L. Wigley, a believer and a well-known climatologist from the University of East Anglia in Norwich, England, has conceded that the warming trend in the southern hemisphere "is clearly inconsistent with a simple interpretation of the greenhouse effect, and what it means is that there are other things going on."

When viewed from a distance, the non-believers seem to make the stronger case. However, they have a profound advantage. They can propose any number of explanations for greenhouse heating. If one explanation is disproven, they can simply forward another. On the other hand, believers must stick by their explanation and attempt to prove it. They must also busily disprove the latest theories advanced by non-believers.

To date, though, believers have the strongest basis for their case. As far as we can determine, temperatures appear to have increased, and the greenhouse effect must eventually cause the atmosphere to warm. Believers have staked their argument on the theory that greenhouse heating has already begun, which may be a mistake.

Again, the real question is: "How much heating can we expect, and when will it occur?" Some calculations project that heating will not be evident for another 100 years, because warming the oceans could take that long. Until the oceans stop absorbing the extra heat from the atmosphere, little atmospheric warming may occur.

If the projection is accurate, the world could be in big trouble. Let's suppose we wait until there is absolutely no doubt that heating has started before we accept the greenhouse effect as real—and that oceans delay the appearance of heating for 100 years. Our failure to act in the interim would mean that we had pumped greenhouse gases into the atmosphere for 100

years too long before admitting that they cause problems. Stopping then might be too late. Temperatures that took 100 years to increase would probably continue to rise for at least another 100 years after we stop injecting greenhouse gases.

SCIENTIFIC CENSORSHIP

Mentioning censorship in our society raises people's hackles. We strongly believe in letting everyone have a say. Generally, the better-educated the person, the more likely we believe he or she is to present all sides of a controversy.

Scientists are well-educated, some say overeducated. You would expect them to object loudly to censorship. They generally do so when political matters are involved, or when someone tries to ban a book. When it comes to scientific matters, however, scientists in effect censor each other all the time. This practice of censorship hits those who hold the minority view—such as greenhouse non-believers—particularly hard. The world of scientific publishing lends itself to easy manipulation by those who hold the majority view.

Most people probably believe that a person who writes a scientific article gets paid for that article. Not so. The scientists, or their institutions, must pay a few hundred dollars for each page published in most American scientific journals. The government funds most of the published research, which means the government pays most of the bills for scientific publishing.

Why are scientists and institutions willing to spend hundreds or thousands of dollars to have an article published? Because publication lends prestige. In collegiate circles, many observe that having a Ph.D. means one must "publish or perish." Scientists build reputations on the length of their publication lists. Universities demand that a professor publish widely before offering him or her tenure in the position. Universities thrive on the prestige that results when their professors publish articles.

Editors of scientific journals have little to say about what appears in the pages of their journals. Not-for-profit scientific societies publish most such journals, and the editor is often a scientist who accepts the unpaid, part-time position for the prestige it affords. The scientific society that publishes the journal establishes criteria that must be satisfied by articles submitted for publication. The fact that a scientist is a member of the society does not mean that he or she has the right to publish in the society's journal.

Because the scientific society wants to use the pages of its journal to best effect. it requires that referees or reviewers must approve an article before publication. The editor generally selects the reviewers, and is pretty much forced to accept the reviewers' recommendations. Reviewers are probably necessary, because one editor cannot possibly understand the details of articles within many specialized technical fields. Lack of sufficient understanding nonetheless occurs, even in journals that specialize in narrow areas like quantum electronics, infrared physics, or pattern recognition.

At any rate, an editor picks reviewers who are specialists in the narrow topic covered by the submitted article. Each reviewer must report whether the article contains errors, and judge whether the article merits space within the journal. Reviewers are free to accept an article, reject it, or require revisions. An article typically goes to three reviewers, all of whom must accept it. Obviously, each reviewer has a significant amount of power.

It is important to realize that the reviewers work anonymously. They can squelch an article without having to face the irate author. Undoubtedly, most reviewers try to provide valid criticisms. But if the article disagrees with that reviewer's deeply felt opinion, the reviewer may prevent its publication. This can effectively censor minority opinions.

Does this ever happen? Of course it does. Consider the case of Richard Lindzen, a scientist who doubts the current greenhouse heating claims.

Lindzen is a highly qualified and respected scientist with degrees from Harvard, and he occupies the Sloane Professor of Meteorology Chair at the Massachusetts Institute of Technology. He is also a member of the National Academy of Science, the nation's most elite collection of scientists.

You might expect a reviewer to say, "Others have a right to see any article written by a person of this stature." This does not hold true in scientific publishing. The *Bulletin of the American Meteorological Society* rejected one of Lindzen's articles, despite the fact that the reviewer was unable to pinpoint one specific error or problem with the piece.

The article dealt with negative feedbacks in the atmosphere and the shortcomings of global circulation models. Lindzen believes that compensating effects in the atmosphere will cancel most—but not all—greenhouse heating. The compensating effects he cites include developments like increased low-level clouds. He expects heating of perhaps as much as 1°F from a doubling of atmospheric carbon dioxide. In other words, Lindzen believes the models omit or underestimate negative feedbacks; he claims the models are wrong.

The reviewer in this case was Jerry D. Mahlman, the director of NOAA's Geophysical Fluid Dynamics Laboratory, the organization responsible for a major global circulation model. Mahlman might be expected to disagree with anyone who attacks his models.

In his review, Mahlman recommended "the paper be rejected unless he [Lindzen] wanted to convert it into a paper about science. It came across as a whiny complaint without scientific justification."

Science, one of the country's largest and most prestigious scientific publications, later ran a story about Lindzen's difficulties. The story revealed the reviewer's name and comments, and the editor of the *Bulletin*—a smaller and more specialized journal—perhaps felt the heat generated by the publicity. A few months later an article by Lindzen, entitled "Some Coolness Concerning Global Warming," appeared in the *Bulletin.*

In the introduction to this article, Lindzen says, "The existence of skepticism on this issue [greenhouse warming] has only recently been publicly recognized." He also refers to "recommendations that skepticism be stifled."

Another article, "Response to Skeptics of Global Warming," which presented the opposing view, appeared later in the *Bulletin.* Will Kellogg, a senior scientist at the National Center for Atmospheric Research in Boulder, Colorado, wrote the article, which supported the notion that greenhouse warming has already started. In his piece, Kellogg called for critical analysis and refutation of statements by those from the "skeptical school of thought."

In the concluding section of his article, Kellogg says, "Thus, any prediction of how the earth will respond to a steadily increasing concentration of greenhouse gases must include the words of caution—it is still too early to be definitive."

If he had stopped there, almost everyone would have agreed with him. However, he could not resist starting the next paragraph with, "Nevertheless, it is even more presumptuous to claim that we do not know enough to 'calculate with confidence' that there will be a significant global warming."

Whether the Lindzen case is a clear example of scientific censorship is a matter of opinion. However, such censorship does exist. Perhaps a means of keeping errors out of the scientific literature is necessary and helps reduce confusion. However, most scientists admit that reviewers come down particularly hard on minority views. As a result, these views may not be heard by other scientists or by journalists covering science beats. And if journalists aren't exposed to a variety of viewpoints, the general public will receive skewed reports on scientific thinking. Thus, censorship often perpetuates misinformation.

THE BIG ISSUE

Frankly, things could be a lot worse. As we discussed at the beginning of this chapter, life will go on even if greenhouse warming amounts to the worst of the current predictions. There is no threat of massive loss of life through rising seas or famines. The proviso here is, if the worst predictions are accurate.

It seems that no one ever mentions a major shortcoming of the model computations with doubled carbon dioxide. The computations assume doubled carbon dioxide and no change in other greenhouse gases. Predictions and the current rate of increase indicate that the combined effects of methane, nitrogen oxides, and CFCs could provide global heating comparable to the contributions from increased carbon dioxide. We know the cause of the increases in carbon dioxide, but considerable uncertainty exists concerning the sources of the other greenhouse gases. Not knowing where they come from makes their increases harder to predict and control.

If the other greenhouse gases continue to increase, greenhouse heating could amount to twice as much as the models indicate. That much heating could, in fact, melt the ice caps and wreak havoc with agricultural production.

The possibility of unpredicted events is undeniably scary. Unsuspected trip points and unrecognized couplings within the global climate system could trigger disaster. The difficult part about warning the world that such disasters might occur lies in finding examples. Any example of a particular problem must involve a known trip point, not an unknown, even though allowing any change in temperatures runs the risk of setting off an unknown trip point. Temperatures are higher than they have been at any time since man evolved (though not higher than at any time in geological history). Allowing still higher temperatures is to blindly stumble about in the unknown.

One chance we take is triggering an ice age. No one knows what tips the balance and starts an ice age. Looking at the temperature history of the world, it is obvious that the weather warms, then starts to cool rapidly. Perhaps, paradoxically, warming is the cause of ice ages. No experts are suggesting this explanation, but experts often operate better in hindsight. The chance may be small, but it is yet another chance. The world is due for another ice age within the next few thousand years, and rushing it would be a shame.

Idso's result, which indicated that increased forest growth could limit future carbon dioxide concentrations to less than double, sounds hopeful. Before continuing "business as usual," though, we must consider the effects of other greenhouse gases. We must also find the will to stop the excess cutting of forests—or at least replace what we cut. We must also realize that trees' ability to absorb the extra carbon dioxide assumes no increase in emissions. If we increase emissions, something different will happen.

Finally, there is the matter of time. We have to plan for life 100 years from now, even though we cannot accurately envision it. If we fail to prepare, the future may turn out worse than expected.

▲ 9 ▼

Ice Ages Come and Go

They could use a little greenhouse heating.

This chapter may set your head spinning. All the things you thought you knew—even those you learned in science class—are not quite true. The reason ice ages come and go, the way we find out how cold it was during ice ages, what an ice age is like, and even the directions in which compasses point—all these depend on truisms from Science 101 that have been, shall we say, overstated.

Here is an easy fallacy to spot: Until they think about it, most people assume that spring and summer contain the same number of days as fall and winter. They're wrong. Winter runs from December 21 to March 20, which is 89 days; spring from March 20 to June 21, which is 93 days. Summer runs from June 21 to September 22, or 93 days; and fall from September 22 to December 21, or 90 days. Spring and summer last a combined total of 186 days, while fall and winter last for only 179 days.

No wonder we have global warming. Somebody put too many days into spring and summer! Of course, the seasons are reversed in Australia, so people there might contend that fall and winter last too long.

Spring and summer are longer in the northern hemisphere because changes of season are physical events, marked by the sun crossing a line or heading south. The difference in the length of seasons does affect the relative temperatures between the northern and southern hemispheres, but other things also contribute to temperature differences.

Many people have the idea that ice ages happened millions of years ago, when awe-inspiring, now-extinct creatures roamed the world. In fact, the most recent ice age ended just 10,000 years ago. Ten thousand years is not that long ago—King Tut lived about 3,000 years ago. More recently, just a couple of hundred years ago, things got cold enough to be called a "little ice age." The little ice age ended at about the time of the American Revolution.

Ice does not cover the whole world during an ice age. It builds up toward the poles, but temperatures in the tropics do not change all that much. Deep ice does cover much of the world, but not the entire world. Try to visualize this: During the worst of the ice age that ended 10,000 years ago, ice was about a mile deep in middle America—in places like Chicago and Detroit. During ice ages, so much ice builds up that sea levels drop by perhaps 300 feet. Spaces that are now oceans become land. We often hear that sea levels dropped low enough during the last ice age to create a land bridge across the Bering Strait, the narrow strip of ocean between Alaska and Siberia. However, with ice cover thousands of feet deep, no one could have known whether there was a land bridge beneath.

In recent years, books predicting imminent doom due to global warming and melting ice caps have become best-sellers. But during the 1970s, a different set of books was selling. At that time, the people who make scientific names for themselves by predicting disasters were warning us about the coming ice age.

Believe it or not, one difficulty that arises when collecting information about global warming is trying to separate the things that may turn the world into an oven from the things that may turn it into a deep-freeze. Some statistics cited as evidence of global warming are also regarded as signs of a coming ice age.

Here is one example. Twenty years ago, scientific authors were publishing pictures demonstrating that Arctic glaciers were moving farther south each year. The photographs served as evidence of the coming ice age. Now, authors are pointing to that same line of ice and snow and saying it indicates global warming, because warmer air holds more water vapor and produces more snow. Perhaps the scientists who foresaw an ice age lived in a different world, or perhaps their ice age lasted just 20 years. In any case, it is amazing how much hotter the people who promote global warming make the world sound now than it was 20 years ago.

Even more amazing is the number of disasters we have apparently averted in the latter half of this century. Foreseen catastrophes started in the 1950s, with the threat

of nuclear annihilation. At the time, many people built fallout shelters and stocked up on food in preparation for the imminent nuclear war.

Since then, we have cut down on the number of children we are having because someone predicted that the population bomb would create massive worldwide starvation. There have been other predictions telling us that we would run out of oil by now and would freeze to death. Even the nuclear winter scares (which contended that high clouds would cool the earth instead of warming it) and the push to buy gold for the coming financial collapse of the world have had their day. Right now, we're preoccupied with the AIDS epidemic, the ozone hole, and global warming, but hints of the next disaster are already materializing—predictions that the world will be decimated by giant meteors. Some scientists claim we need to build bigger-than-ever nuclear bombs so we can disintegrate those incoming meteors—and even the death of dinosaurs is now blamed on the impact of a meteor. It sometimes seems that the world has more problems than the heroine of a soap opera.

We can learn a serious and valuable lesson from the onslaught of disaster scenarios: healthy skepticism. No matter what the prediction, nothing is gained from panic, hopelessness, or blind belief. When an authority figure announces that the sky is falling, we can respond with intelligent questions:

"Which way?"

"How soon?"

And most importantly:

"How do you know?"

DISCOVERY AND EXPLANATION OF ICE AGES

In the 1830s, geology was a new subject. No one had conceived of ice ages at the time, so no one realized that a so-called "little ice age" had just ended. Some evidence suggested that the world had once been covered with ice, but the phenomenon was explained differently. At that time geologists accepted catastrophe theory—the concept that the world had undergone a series of catastrophes called epochs. Each epoch ended with a catastrophe that wiped all life from the earth. The most recent was believed to have been the flood, for which Noah had constructed his ark.

This theory was easily capable of answering difficult questions. If someone presented an ancient fossil to a geologist, he or she was told it came from a strange creature that had lived in an earlier epoch. The theory allowed us to keep the Biblical explanation of Creation intact, but still accommodated the existence of weird creatures in the distant past.

Sometimes strange reasoning created even stranger creatures, such as a human with huge teeth. In their book *Ice Ages,* John and Katherine Imbrie recount a letter from Governor Dudley of Massachusetts to Cotton Mather, the Boston preacher. The letter concerns a fossilized mastodon tooth that was unearthed in New York in 1706:

"I suppose all the surgeons in town have seen it, and I am perfectly of the opinion that it was a human tooth. I measured it, and as it stood upright it was six inches high lacking one eight, and round 13 inches, lacking one eight, and its weight in the scale was two pounds and four ounces, Troy weight . . . I am perfectly of the opinion that the tooth will agree only to a human body, for whom the flood only could prepare a funeral; and without doubt he waded as long as he could keep his head above the water, but must at length be confounded with all other creatures and the new sediment after the flood gave him the depth we now find."

Scientists used this kind of reasoning when large rocks were found, which had been carried far from their original bedrock by glaciers. The displaced rocks, as well as the glacial scratching and polishing marks upon them, were attributed to movement by the Biblical flood. This even explained rocks high in the mountains—after all, the flood had covered the whole world.

Louis Agassiz was a charismatic Swiss scientist who upset the world of geology in the 1830s. He aggressively promoted the idea that ice once covered northern Europe. Others had conceived the idea before him, but Agassiz disputed the conservative geologists of the time and popularized the theory that an ice age had occurred. He offered no good explanation for what caused the ice age, perhaps because he had enough trouble achieving acceptance for the ice age even when it wasn't burdened with theory and explanations.

When a new discovery occurs, dozens of scientists generally devote themselves to explaining how the discovery works. This happened with ice ages theory. For 25 years, a battle raged about whether ice had once covered the world. The idea that so much ice could have existed, some experts claimed, was ridiculous. In my experience, experts often oppose new theories because they are expert only in old theories, not new ones.

When the existence of the ice was finally accepted, a host of wild-eyed theories were offered as explanations. For example, one theory held that the land on earth had suddenly been elevated to higher altitudes, where it is colder. The colder temperatures at higher altitudes caused ice sheets to form. Why the land might have been subject to sudden elevation was not explained.

There were theories claiming that the sun's heat had decreased, and these theories exist today. Many people suggested that smoke from volcanoes caused a decrease in the solar energy reaching the earth, creating a sort "volcano winter" in lieu of a nuclear winter.

Another set of theories suggested that changes in the earth's rotation caused the ice ages. As the earth rotates on its own axis once each day and around the sun once each year, strange things do happen. The earth's circuit around the sun takes place in a flat plane, as shown in Figure 1. If that circuit were a circle, and if the earth's own axis was perpendicular to that plane, everything would be simple. We would not experience winter or summer. Every day would offer equal hours of sunshine and darkness. However, the earth's axis falls short of being perpendicular by 23.5 degrees. As the figure shows, this angle causes one pole to enjoy six months of sunshine while the other is enduring six months of night. The tilt means that sunshine only reaches both poles simultaneously twice a year, during the spring and fall equinoxes. Also, the circuit the earth makes around the sun is an ellipse with the sun located at one focus of the ellipse. It is not truly a circle with the sun at the center. As a result, seasons differ in length.

The figure also shows the equinoxes and solstices that define the seasons, and the greatest and least distances from the sun (aphelion and perihelion). The equinoxes occur on the days when the sun is directly over the equator. On these dates, the angle between the earth's axis and the direction of the sun is a right angle. On the equinoxes, day and night are the same length. The solstices occur on the dates when the sun reaches its greatest distance from the equator, the longest and shortest days of the year.

Earth's orbit is only slightly elliptical. The figure exaggerates how elliptical the earth's orbit is, so we can clearly see that the sun is at a focus instead of at the center of the ellipse.

As Figure 1 shows, the tip of the earth's axis causes the sun to shine for more than 12 hours a day during summer and less than 12 hours a day during winter. Thus, the tip of the axis essentially causes summer and winter. Another factor is the fact that sunshine strikes the ground at a more oblique angle during winter, while it shines down more directly during summer. At the poles, the earth's tipped axis means the sun never sets on the long days of summer and never shines on the short days of winter.

As strange as it may sound, the angular

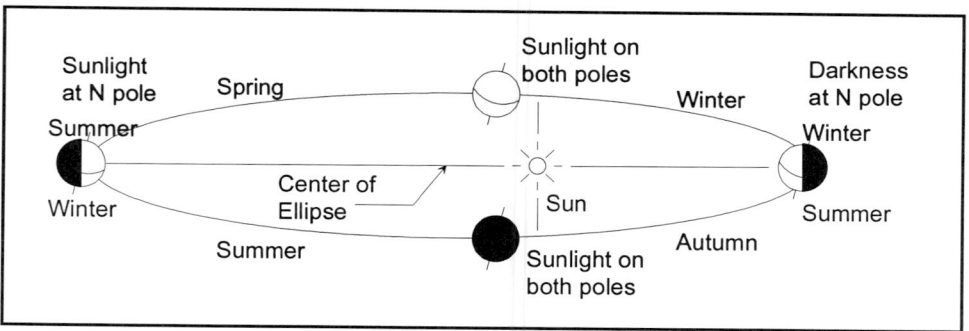

FIGURE 1 *Earth makes an elliptical trip around the sun each year. The tip in the earth's axis—not the distance from the sun—causes the seasons. Longer days in summer mean more solar heating. The sun is at a focal point of the ellipse, but is not its center, which makes northern hemisphere spring and summer longer than fall and winter. Sunshine reaches both poles on the same day only at the beginning of spring and autumn.*

tipping of the earth's axis keeps changing. Over a 41,000-year cycle, the tilt of the earth's axis varies from 21.8 to 24.4 degrees. At the greater tilts, differences between winter and summer become more pronounced.

Earth's elliptical orbit means the distance between the sun and the earth varies from 91 to 94 million miles during the year. The sun is actually closer during northern hemisphere winter than it is during northern hemisphere summer. Notice that the tilt—not the distance from the sun—determines winter and summer. It is not the case that moving farther away from the sun makes no difference. Rather, the tilt has a stronger effect than the distance. An increase of three million miles in the distance between the earth and the sun decreases the earth's solar heating by about six percent. This decrease is less than the combined effects of longer days and the more direct angle of sunlight in summer.

Earth's orbit changes, too. That may also sound strange and wrong; perhaps the earth wobbles on its axis as it spins, but what would cause the earth's path around the sun to change? Other planets cause the change. As the earth rotates around the sun, it is swayed by the gravitational attraction of other planets. That attraction causes the orbit to change, but it repeats itself every 100,000 years. Naturally, the variation in the orbit affects how much heat we get from the sun.

Earth's wobble on its axis is given the more scientific name of "precession." In everyday experience, it corresponds to the way the axis of a child's top changes direction as the top rotates, as shown in Figure 2.

Precession is very slow. A complete cycle requires 23,000 years. Notice that, as the axis precesses, it points in a different direction. The North Pole is always on the rotation axis, because the earth keeps

rotating about the same axis. However, a change occurs in the direction in which the axis points. Precession dictates that the axis will start pointing in a different direction, and toward different stars. The axis now points at the North Star, toward the end of the handle of the Little Dipper. In 2000 B.C., however, the axis pointed toward a spot midway between the Big Dipper and the Little Dipper. In 4000 B.C., the North Pole pointed toward the tip of the Big Dipper's handle.

People have proposed one or the other of these orbital effects as the cause of ice ages. In fact, orbital theories have been popular in explaining ice ages about as often as wide neckties have been in style

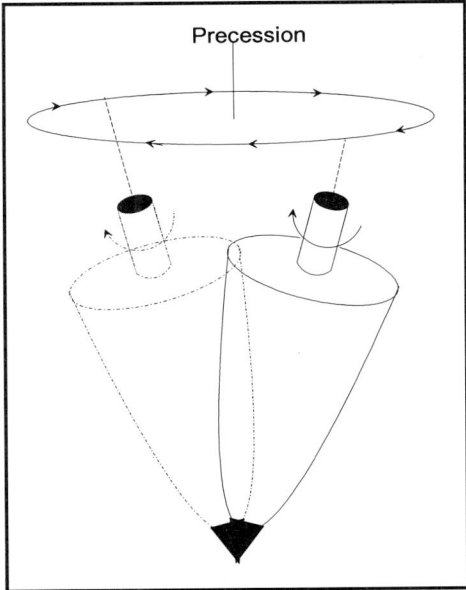

FIGURE 2 *As a top spins, its axis slowly precesses. Earth is like a big top, so its axis also precesses as it spins. This precession affects the direction in which the axis points, and the points in the orbit where the seasons change. Right now, the earth is farther from the sun during northern hemisphere summer, but precession causes that to vary. Being farther from the sun in summer decreases heating by about one percent.*

for men. At about the time of World War I, a fellow by the name of Milankovitch proposed that ice ages are caused by a combination of three orbital effects. He even calculated the effects of the three on sunshine in different north-south zones. According to Milankovitch, wobble, change of axial tilt, and modifications of the elliptical orbit jointly cause ice ages. Today, most scientists accept the Milankovitch theory, but some insist that it miscalculates the dates during which ice ages occurred. Oceanographers tend to favor the theory because they claim that data from the ocean bed confirms Milankovitch's dates.

Scientists turn their nose up at the very mention of astrology, which contends that the position of the planets determines what happens to people on earth. Any scientist will assure you that there is no connection between the minute forces that the planets exert on the earth and what happens here. Yet these same scientists believe the Milankovitch theory, which holds that planets change the earth's orbit and trigger ice ages.

HISTORY OF THE ICE AGES

Milankovitch's theory forwards one overriding argument that led to its acceptance. It predicts that ice ages occur on a regular timetable, and this timetable seems to agree with the periods during which past ice ages happened. The timetable is critical to the Milankovitch theory, so let's examine the long-term temperature history of the world, shown in Figure 3.

Once we are told that ice ages occur every 100,000 years, the 100,000-year period looks somewhat obvious. If we're not told, however, the cycle is less obvious. For example, the ice age about 500,000 years ago was pretty weak—not

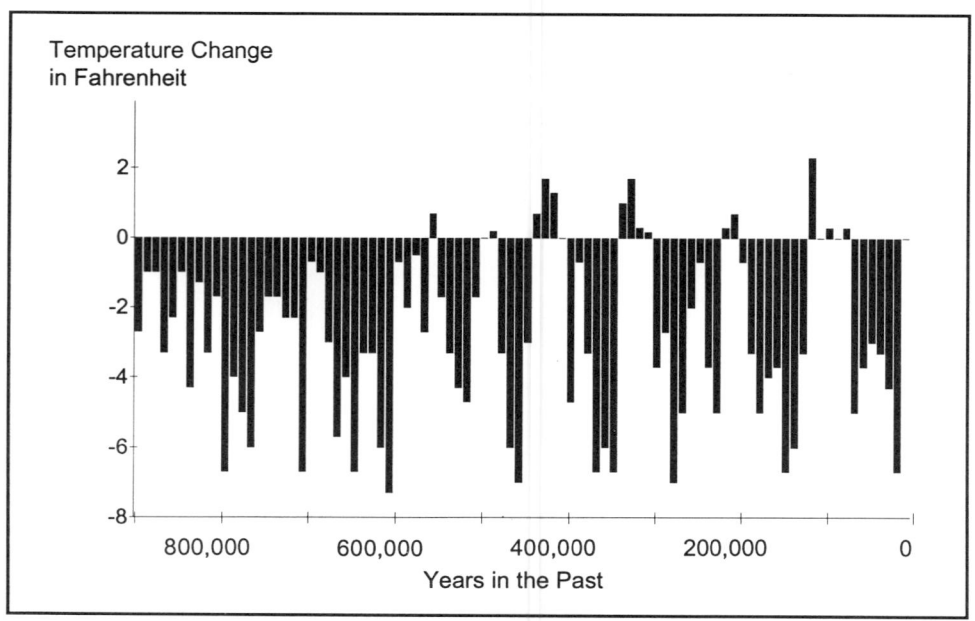

FIGURE 3 *Ice ages have come and gone over the last million years, without any help from man. Here, you can see that ice ages occur regularly at 100,000-year intervals. Zero temperature change represents our current temperatures. The earth has been warmer several times in the past than it is currently.*

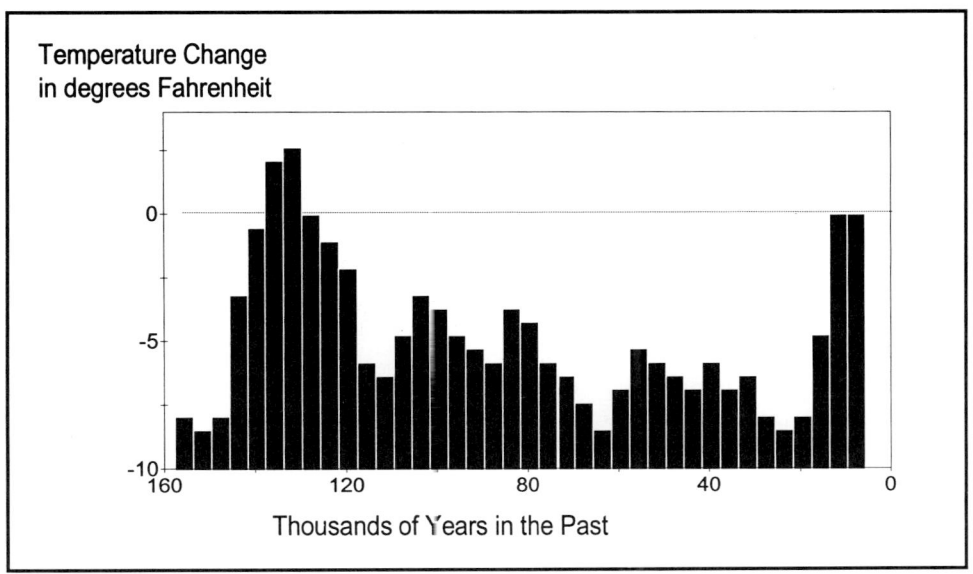

Temperature Change
in degrees Fahrenheit

Thousands of Years in the Past

FIGURE 4 *Temperatures are known over the past 160,000 years. There is more disagreement over the dates at which temperatures occurred than over the temperatures themselves. Notice the scant temperature differences between current levels and ice ages.*

really a "good" ice age. Between 600,000 and 800,000 years ago, it seems that four or five ice ages occurred instead of two or three. The "zero" line in the figure shows the current global average temperature of the world. Notice that at several times in the past, temperatures have been higher than they are at present. In fact, greater temperatures than current levels occur after most ice ages. Proponents of greenhouse heating say the world is hotter now than it has ever been, but that is not so.

The world seems to experience eras called interglacial periods, then suddenly takes a nosedive into an ice age. Ice ages last a relatively short time. As soon as the world gets very cold, temperatures shoot up rapidly and the ice age ends. This rapid cooling and heating contrasts sharply with other, slower temperature changes.

If the only things causing the ice ages were changes in the earth's orbit and rotation, temperatures would slowly oscillate between hot and cold; they wouldn't

change suddenly. Judging from the temperature history, when the world gets down to a certain temperature, a sudden change occurs. The world immediately starts to warm up, and the ice age soon ends. It seems that when the global temperature gets only a few degrees colder than it is at present, sudden heating is triggered.

Notice the temperatures in the accompanying graphs. Temperature differences between the coldest part of the ice ages and the warmest temperatures between ice ages amount to only about 10°F. That is an alarmingly small difference.

It is often said that the warming during this century has taken place more rapidly than ever before in history. Again, this is not so. Taken over 50-year periods, the warming that took place 150,000 years ago (after the preceding ice age) was just as rapid as our current warming. This is indicated in Figure 4, the history of temperatures for the last 150,000 years. The limitation in the warming rate shown by these curves has more to do with past limitations on the measurement of rapid temperature increases than with the actual temperatures. New data indicates that even more rapid increases took place.

Some of the new data comes from Sigfus Johnsen of the University of Copenhagen in Denmark, and from a European team. They obtained ice cores as part of the Greenland Ice Sheet Project (GRIP). Ice initially obtained from these cores goes back 40,000 years and reveals what are called interstadials. These are periods lasting from 500 to 2,000 years, during which temperatures shot up by 12°F. When previously seen in other ice core data, such interstadials were discounted and blamed on ice flow. GRIP is drilling through the thickest part of the Greenland ice sheet. The project has obtained an ice core almost two miles long that goes back about 200,000 years. Their

findings concerning the interstadials emerged after they had studied only the most recent 40,000 years of data.

Some evidence of such short-lived warm spells has also been seen in ocean sediments. So far, the best explanation offered is that surface currents in the North Atlantic changed rapidly. There is no understanding, however, of what might trigger such rapid shifts. A shift could occur at any time, but what might make it shift and then return quickly to its original position remains a mystery.

Johnsen joins others in warning about shifts in ocean currents. "If the current system of the oceans starts to change," he says, "we could have drastic changes in climate."

Since we are at the peak of the warming trend between ice ages, a shift in the currents would probably trigger cooling, rather than heating. One major concern

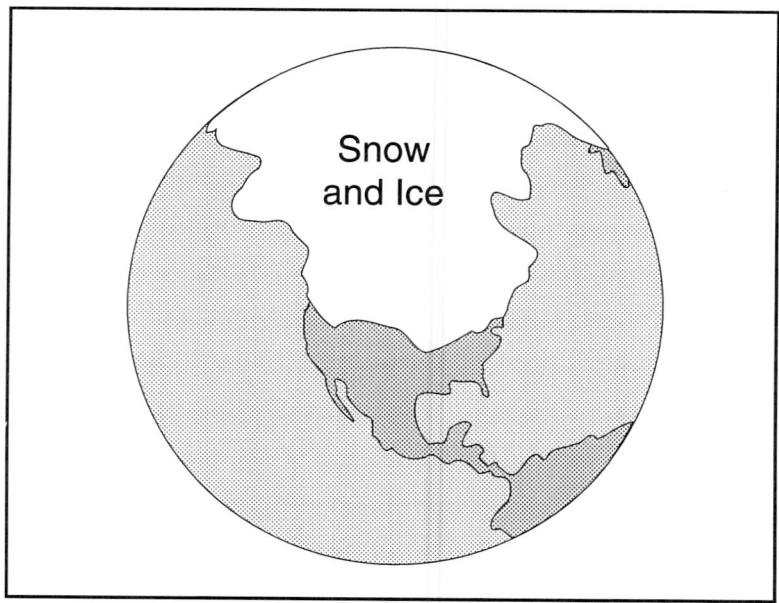

FIGURE 5 *The map shows one set of estimates of what parts of North America were continuously covered with ice during the last ice age. Notice that some ocean areas, like the Bering Strait, became frozen. Freezing the ocean at levels deep enough to cut off water circulation blocks ocean currents and affects where oceans redistribute heat that is absorbed in tropical regions.*

relevant to global warming is that it might somehow throw the switch that sends us into an ice age. So we are confronted with unlikely and opposite worries—that global warming may occur, and that the warming may trigger an ice age. In the past, ice ages have occurred just after the world reached a certain temperature. Global warming could hasten our arrival at that critical temperature.

Civilization can cope with a heat increase of a few degrees more easily than it can cope with another ice age. With warmer temperatures, large areas of the world would become agriculturally useful. Cold temperatures, on the other hand, would shut down farming in Canada, parts of the northern United States, and northern Europe. We can get an idea of just how bad things might become by looking at Figure 5, which shows the glacial coverage of North America during the last ice age.

Much remains unsaid concerning the history of ice ages. Not everyone agrees that the Milankovitch theory can predict when ice ages are coming, for example. A curve of temperature history may make readers believe that scientists know those temperatures with certainty, but this is not so. Some scientists think the time scale is off. Virtually everyone admits that there may be large errors in the temperatures. Remember, part of the argument about whether the world has heated up during this century has to do with how accurate our turn-of-the-century measurements were. Then think of the degree of uncertainty there must be about temperatures measured indirectly from thousands of years ago. To appreciate just how ignorant we are about the history of the ice ages, consider the tremendous problems. Ask, for example, how scientists managed to reconstruct worldwide temperatures at dates that occurred a few hundred thousand years before the invention of the thermometer.

MEASURING PAST TEMPERATURES

Several techniques exist for measuring temperatures that occurred in the distant past. All of them depend on some form of isotopic measurements—more specifically, on small differences between the reactions of different isotopes.

Oxygen is a good example. Both oxygen-16 and oxygen-18 are stable, and do not undergo radioactive decay. The "16" and "18" identifiers indicate the respective weights of the isotopes. Oxygen-16 has the weight of 16 nucleons (eight protons and eight neutrons), while oxygen-18 carries the weight of two more neutrons.

The extra weight slows oxygen-18 down a little. At cooler temperatures, the slower motion of the oxygen-18 means that a smaller fraction of the isotope will leave the ocean through evaporation. Oxygen-18 is also less likely to return to the ground in the form of the ice and snow that remain trapped in glacial ice. This means that, as temperatures cool, glaciers start evidencing a smaller ratio of oxygen-18 to oxygen-16. Therefore, the ratio of oxygen-18 to oxygen-16 depends on the temperature at which the snow formed. Measure the isotopic ratio, and you have a measure of the air temperature at the time when the snow formed. Of course, there are scenarios that can mess up this whole process. Scientists debate them, but this is the most understandable scenario.

Parts of ice caps in Greenland and the Antarctic never melt—not even in the time between ice ages. They build up and get thicker, year after year. All a researcher needs to do is drill a core from the ice caps and measure the ratio of oxygen-18 to oxygen-16. That ratio gives a measure of the temperature at which the ice formed. It does not tell us when the ice was formed, but it does tell us how cold it was.

Oceanographers participate in temperature measurements as well, although they have no layers of ice to work with. Instead, they use layers of calcium carbonate, which contains oxygen. The carbonate forms when small sea creatures die and their skeletons settle to the bottom of the ocean. Ocean temperatures determine the rate at which the sea creatures, prior to their death, incorporated the different oxygen isotopes into their systems. The same isotope ratio is present when they form layers of calcium carbonate. However, there is one complication: During ice ages, glaciers affected the amount of oxygen-18 in ocean water—not much, but a little. Dealing with sea water that contained excess oxygen-18 caused interpretation problems for a time, but these problems have apparently been resolved.

Let's compare ocean measurements and ice measurements. Ice records go back only about 200,000 years. That period of time spans two ice ages and utilizes all the dependable ice we have. Ocean sediments, on the other hand, go back at least a million years. Thus, short-term temperatures, as much as 200,000 years old, can be determined through either ice or ocean measurements. For the really old measurements, we depend on the oceans.

Ultimately, temperatures without age are about as interesting as basketball scores without team names. Oceanographers do have a means of calculating dates—a scheme that depends on the earth's nodding in its revolutions and periodically reversing its magnetic pole. From time to time the north and south magnetic poles reverse places. This does not mean the world flips over; it means that the electrical currents causing the earth's magnetic field change the direction of their flow. We are not sure why magnetic fields reverse, but the changes show up in volcanic lava flows and other rocks. Oceanographers use this phenomenon as part of their scheme for dating pre-historic

temperatures. Other scientists are more skeptical.

Isaac Winograd of the U.S. Geological Survey in Reston, Virginia, is a man with a different calcium carbonate record—one that did not come from the ocean. His measurements came from a 150-foot-deep, water-filled crevice called Devil's Hole, located 80 miles northwest of Las Vegas, Nevada. His carbonate record covers 600,000 years, and it disagrees with what the oceanographers have told us. It presents different dates for the ice ages, and it differs from the oceanographic records in two additional ways.

Winograd's record is taken from air temperatures in Nevada, while other temperature records are based on glacial ice. Even more importantly, Winograd has a better method for dating his carbonate record. He uses chemical species produced by the radioactive decay of uranium—a very reliable form of dating.

On a large scale, Winograd's findings are not greatly different from those of the oceanographers. He indicates that the previous interglacial period ended at least 147,000 years ago—20,000 years sooner than the oceanographers say it did. What is a 20,000-year discrepancy over the course of 147,000 years? Not much, if you are not committed to the Milankovitch cycle of ice ages. If you do feel committed, Winograd's date casts doubt upon "your" (Milankovitch's) theory.

Figure 6 shows a comparison of the isotope ratios provided by oceanographers and those taken from Devil's Hole, along with the solar heating rate according to Milankovitch's theory. Notice that the oceanographer's isotope ratio shows that the next to last ice age ended when solar heating was at one of its 10,000-year highs. Devil's Hole data suggests that the ice age ended earlier, when the solar heating was at a minimum. As supporters of

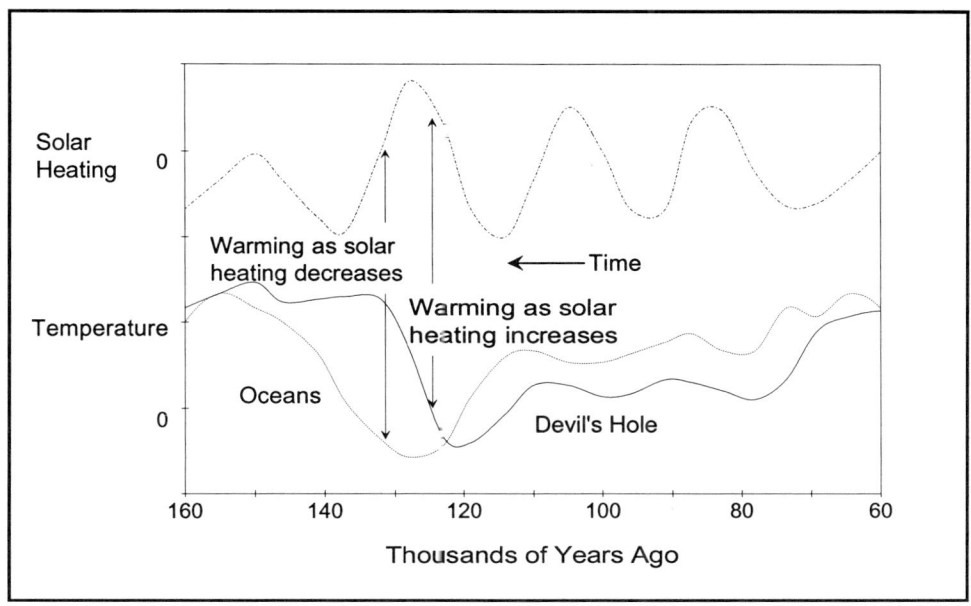

FIGURE 6   *The dates and temperatures obtained from ocean sediments disagree with dates and temperatures from Devil's Hole, Nevada. Devil's Hole temperatures cast doubt on the theory that ice ages are caused by wobble and precession of the earth's axis. Ocean data shows the ice age ending at the peak of solar heating. Devil's Hole data shows the ice age ending after solar heating, while solar heating is decreasing. Solar heating is calculated by considering the changes in the angular direction of the earth's axis.*

Winograd ask, "How could rapid warming come so long before the cause of the heating?" The cause of the heating is a peak in the solar heating rate that lasts for 10,000 years and is due to the earth's wobble.

Signs of a compromise are beginning to appear. Oceanographers now admit that they have long known that the ocean surface in the southern hemisphere warms for several thousand years before ice sheets start shrinking. Other oceanographers say they feel comfortable with the idea that Nevada might simply have undergone a different climate. They say a regional climate could change well before the large-scale melting of the ice sheets reflected in the ice volume signal.

Other oceanographers take the matter straight to heart. William Ruddiman of Columbia University's Lamont-Doherty Geological Observatory is one. He defends the oceanographer's calculations by saying, "Everything fits together so well that it would have to be a preposterously cruel joke if we were wrong."

Time will tell whether nature pulled such a joke, but Ruddiman should understand that Mother Nature never promised to make science simple for the convenience of scientists.

It should be noted that scientists accepted the Milankovitch theory mainly because of its mathematical appeal. The theory was accepted before radioactive carbon dating appeared on the scene. Radiocarbon dating became popular by the 1950s, and it gave the Milankovitch theory a rude jolt. Radiocarbon dating of Canadian peat bogs indicated that the bogs were 25,000 years old. However, peat bogs

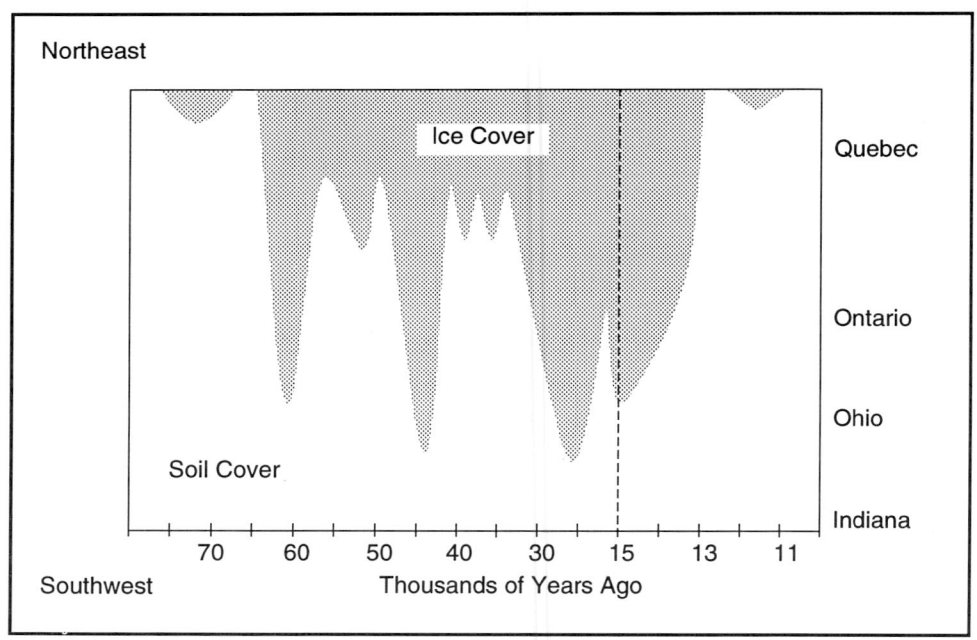

FIGURE 7 *Over time, the north-south limit of constant ice cover has varied from Quebec to northern Indiana. This figure shows how it moved back and forth between the locations on the right over time. Notice the compressed time scale beyond 15,000 years ago. This data came from radiocarbon dating of plant matter.*

do not form in the middle of glaciers, and glaciers should have been occupying the middle of Canada 25,000 years ago. Plant growth was also found in other places, where cold temperatures made plant growth doubtful. As a result, many geologists decided Milankovitch's theory was pretty, but wrong. In 1965, R.P. Godthwait and others published the data in Figure 7. It shows the position of the glacial margin in North America over the last 70,000 years. This data was never disproved, but it disagrees profoundly with Milankovitch.

As if to confuse the picture, dating with ocean sediment yielded answers consistent with Milankovitch, despite some lingering questions about the time calibration. Scientists again accepted Milankovitch, and simply avoided trying to explain how peat bogs could occur at a time and place where glaciers were hundreds of feet thick.

It seems that scientists must have some operating theory. Without theory, there is no context for data. If data seems to disprove the only two existing theories, scientists never cast both theories out. They accept the most elegant theory, or the one that conflicts the least with the data.

Milankovitch's theory came back into vogue only after oceanographers used sea sediment to work out a temperature history. The oceanographers got a history that fit so well with Milankovitch's theory that the theory was resurrected. Of course, the scientists knew the theory and the answers they wanted before they started looking for verification. In my experience, you get a lot of mumbled answers if you ask about the conflicting evidence offered by radiocarbon dating as practiced in Canadian peat bogs.

ICE AGES INVOLVE
MORE THAN ORBITS

In courses on the philosophy of science, instructors teach about Occam's Razor. Occam probably never heard of the rule that bears his name, but scientists tout it as a guiding principle of science. Occam's Razor says: "When two theories are capable of explaining the observed data, the simpler one should be used." An addendum to the rule is necessary, in my estimation. We should add, "provided the explanation is not *too* simple."

Ice ages are a case in point. Why not attribute the ice ages to changes in the output of the sun? It has been suggested that ice ages occurred because the heat radiated by the sun decreased. Another theory suggested that something in space reduced the amount of heat reaching the earth. These "reduction of heating" theories never got much consideration. They were too simple and too obvious.

When scientists first sought an explanation for the ice ages, no one knew what made the sun hot. No one knew that nuclear fusion reactions drive the sun. An explanation saying the output of the sun had decreased offered no opportunity for scientific argument. It was hard to develop any interest in such a theory, because it was not verifiable. Scientists want to take measurements, make observations, and develop astute arguments. Recently, there have been suggestions that global warming may be due to increases in solar output. These suggestions have generated some interest. We now have years of observations from the ground and from satellites to sift through with computers, in order to try to find a trend of variations.

If you look hard enough, you can find almost anything. A variation in solar output has in fact been reported. The variation amounted to just 0.04°F during the 1980s. That is too small to explain global warm-

ing. Still, a few people claim an even smaller solar effect—sunspots—causes the ice ages. Sunspots are storms on the sun that observers see as small, darker spots on the solar surface. The amount of interest in sunspots over the years is amazing. Records exist that give the numbers of sunspots for the last 300 years.

Sunspot activity is periodic, but not perfectly so. During a cycle, which lasts about 11 years, the number of sunspots increases from almost none to a maximum number, then declines again. A history of the sunspot numbers is shown in Figure 8. The cycle repeats at regular intervals, but during some cycles more sunspots occur than during others.

Scientists, crackpots, and other observers have claimed that a relationship existed between sunspots and almost anything you can imagine. One theory, which claimed that sunspots reflected the price of stocks on the stock market, received much attention but created few millionaires. Correlations can be drawn between sunspots and tree rings, sunspots and volcanic activity, and sunspots and ice ages (there were fewer sunspots at the start of the little ice age). Others say that ozone levels vary in phase with sunspots. No significant variation in heat from the sun seems to accompany sunspots, so the physical processes that tie the world's destiny to sunspots have never been clear. People keep finding statistical connections, but until someone establishes a cause and effect relationship, sunspots will remain nothing more than a curious phenomenon that may exert some control over our lives in some mysterious manner.

CLIMATIC CONNECTIONS

Meteorologists are suspicious of most explanations for ice ages. As experts in weather forecasting, they like to believe that all they need in order to predict the

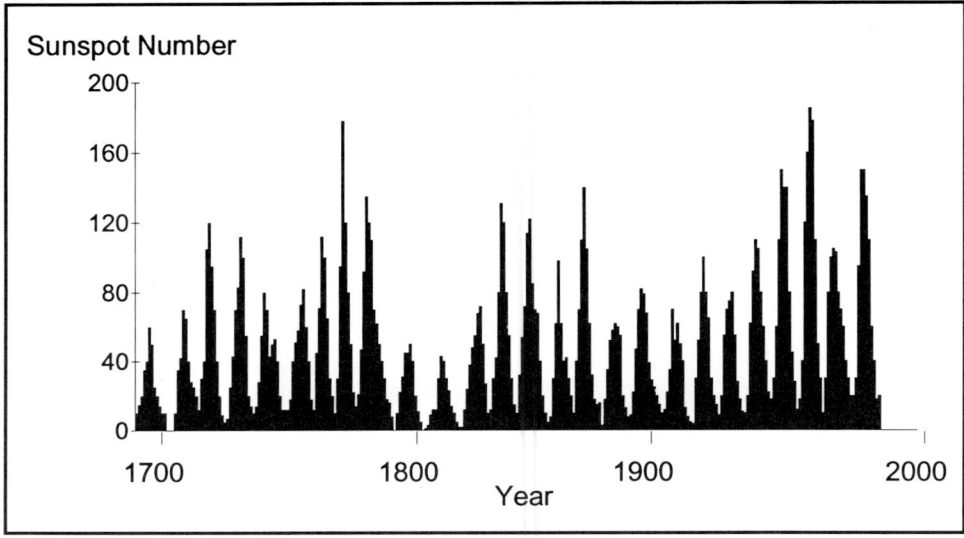

FIGURE 8 *Sunspots may or may not affect our climate. Many theories have tried to relate the two, but the tie-ins were generally not accepted. Perhaps the most interesting thing about sunspots is that such a long history of observation is available. Sunspot history goes back much further than reliable measurements of things here on earth, like temperature and rainfall.*

weather is a worldwide map of temperature, pressure, and humidity. If you tell them the biggest weather events over the millennia depend on some small effect like the number of sunspots or slight variations in the earth's rotation, they get uncomfortable. Saying such things is really saying that meteorologists cannot predict the granddaddy of all cold spells.

The wobbling of the earth's axis is a cycle that takes thousands of years. To the meteorologist, such a small effect over such a long period seems trivial. After all, the total amount of heat the world receives from the sun is unchanged, no matter how much the earth wobbles. Perhaps a little more goes to the southern hemisphere when the world is tipped, but only one percent or so. During thousands of years, weather and ocean currents should keep that small energy difference well-mixed between the hemispheres.

Meteorologists have a good argument. It seems reasonable that the average tem-perature of the world should depend on how much heat we get from the sun. Tipping the earth does not change the total heat, only where the heat falls.

James Croll, a Scotsman, offered a rebuttal against that argument in the middle of the last century. Croll was the first to connect the orbital eccentricities of the earth and ice ages, well before Milankovitch. According to historical records, he had a "modest, shy, dry, and almost speechless manner." He was a self-educated man who was finally recognized for his work and raised from a position as janitor to become a Fellow of the Royal Society.

According to Croll, decreasing the solar heating in one hemisphere causes a positive feedback, which diverts even more heat away from that hemisphere. The connections are ocean currents and trade winds. Trade winds are the strong winds, important in the days of sailing ships, that blow along the equatorial regions. They provide a major driving force for ocean surface currents. If one hemisphere became colder, according to Croll, more north-south heat transport would occur and cause an increase in the trade winds. That increase in winds would affect the ocean currents, particularly the rotational flow in the South Atlantic.

The counterclockwise-flowing gyre that circulates water around the South Atlantic ocean brings a warm current toward the eastern tip of Brazil. Croll recognized that this land tip extends far into the ocean, deflecting part of the warm current from the southern hemisphere into the northern hemisphere. Heat carried in this deflected water helps warm the northern hemisphere and keep the hemispheres in balance.

The deflected warm water also helps prevent ice ages. Ice ages happen when less warm water is deflected north. Here is a summary of Croll's description of how the earth's wobble could cause an ice age:

Suppose the earth's axis tips a little, so the northern hemisphere gets less heat. Temperatures at the equator would hardly change, but the decrease in the already sun-short northern latitudes would result in significant cooling. This cooling would mean a steeper temperature gradient across the northern hemisphere, causing trade winds to blow harder. Stronger trade winds blowing along the northern side of the equator would increase flow in the clockwise-flowing North Atlantic gyre. This would shift the place where the warm South Atlantic gyre hits the tip of Brazil. No longer would Brazil's sharp tip deflect as much of the warm southern hemisphere flow into the northern hemisphere. Decreased flow of warm water from the southern hemisphere, in turn, would cause the northern hemisphere to cool even more.

In essence, the process as Croll envisions it initiates a feedback that intensifies the temperature differences between the hemispheres. It sounds reasonable, but a curious sort of ice age would result. The feedback mechanism would cause the northern hemisphere to get colder and the southern hemisphere to get warmer. The southern hemisphere would heat up because it loses less warm water to the north and receives less cold water from the north. The northern hemisphere might freeze up, but the southern hemisphere would become positively toasty. It would be a great time to head south.

But ice ages do not work that way. Both hemispheres become cold at the same time, so moving to the southern hemisphere would not solve your problem if an ice age began. Remember the meteorologists' objection. If the same amount of solar heat reaches the world, how could small changes in its distribution drive both hemispheres into an ice age at the same time?

First and foremost, the two hemispheres are not the same. The southern hemisphere contains less land than the northern hemisphere. It has cooler air temperatures, but warmer water temperatures, a dichotomy that results from the ratio of land to ocean. Oceans are almost black, and they absorb as much as 90 percent of the sunlight that strikes them. Land and plants reflect and scatter a larger fraction of the sunlight. So the predominance of oceans "down under" means that the southern hemisphere captures more of the available solar heat.

Air absorbs little sunlight. Instead, the sun heats the ground surface, and the surface then heats the air. Sunlight shining on the ground can cause it to become very hot, as you know if you have walked barefoot across a sandy beach on a sunny day. However, the water at the edge of the beach is much cooler than the sand. Water mixes the heat from the sun with some of the water below its surface, whereas heat absorbed at the top of the sand remains unmixed. The heat stays on top.

It is clear, then, that air in contact with

water is heated less than air in contact with earth. And since the southern hemisphere has less land than the northern hemisphere, air in the southern portion of the world is cooler than air in the northern hemisphere. More heat is absorbed in the southern hemisphere because of the dark oceans, but less is transferred to the air. Conversely, water is heated to a greater degree in the southern hemisphere. The tip of Brazil deflects some of that water to the north, keeping the northern hemisphere warmer.

Feedback mechanisms like snow and ice come into play here. As a hemisphere cools, snow cover increases. Oceans absorb almost all the sunlight striking them; snow absorbs almost none. Suppose the weather cools, and snow increases. Increased snow cover reflects more solar heat, so the weather cools further. The additional cooling may not trigger more snow, but it does ensure that the snow already fallen will remain on the ground for a longer period of time. The additional cooling might also trigger snow in locations that are normally snow-free.

Once a cooling process is underway, other mechanisms may kick in. The water to generate all that snow and ice has to come from somewhere, and that somewhere is the oceans. During ice ages, sea level drops because so much ice is deposited on the land, and there is only so much water in the world. During the last ice age, sea levels fell by 300 to 400 feet. When sea level falls that dramatically, ocean currents shift. Water no longer flows through regions like the Bering Strait or the English Channel. With a 350-foot drop in sea level, both would become land rather than sea. They might be covered with ice, but it would not be frozen sea ice.

Such a distinction is important, because the ocean water that usually flows beneath the surface of frozen seas would no longer flow in some areas, and this is the flow that transports heat and helps thaw the ice cover in spring and summer. This brings to mind a major difference between Arctic and Antarctic ice. There is no land at the North Pole, so water can flow under Arctic ice and help it melt. However, the Antarctic continent keeps the flow of ocean water well away from the South Pole, which is why ice fields at the South Pole are less likely to melt than those at the North Pole.

These feedbacks all intensify cooling. They explain why temperatures drop so quickly when an ice age has begun. But of course another question is raised: What finally brings the world out of an ice age? Ice ages obviously do not last forever, and temperatures rise more rapidly at the end of an ice age than they fell when the ice age started. It seems a mysterious reversal.

Most warming feedbacks are simply the opposite of the cooling feedbacks that drive the world into ice ages. As an example, consider snow. At the beginning of the ice age, cooling ensured that snow would last longer, and the reflectivity of the snow caused decreases in solar heating. This amounted to a cooling feedback. Once warming starts, snow cover decreases. More sunlight is absorbed, leading to further warming. The melting of the snow is a warming feedback.

Feedbacks operate in both directions. Positive feedbacks drive the weather further in the direction it is already heading. However, no one can sufficiently explain the changes in climatological direction. We do not know what causes a change from warming to an ice age, or from an ice age back to warming. It could be a reflection of the Milankovitch cycle, or it could be an effect that remains undiscovered.

GREENHOUSE GASES
AND ICE AGES

As we've seen, the current concern about global warming centers around greenhouse gases. Concentrations of these gases have varied over time, without any help from humans. "If they change on their own," someone might well ask, "what's the difference if humans change them, too?" But the greenhouse gas concentrations do not change on their own. Something changes them, and a better question is this: Can we can look at the changes provided by nature and use these to predict what might happen if humans cause the same changes?

Scientists use ice cores to obtain samples of gas from past atmospheres. The ice has small bubbles in it, just like the bubbles that make ice cubes white instead of clear. Crushing the ice cores taken from ice caps breaks the bubbles and releases a sample of the long-entrapped air.

The results confirm the predictions of greenhouse theory. During interglacial periods, concentrations of carbon dioxide and methane are high. During ice ages, concentrations of both greenhouse gases are low. Superficially, it appears to be an open and shut case. The cause and effect relation seems obvious. But perhaps we should look beneath the obvious.

Is it also possible that decreased temperatures might be the cause of, not the effect of, diminished greenhouse gases? This certainly happens when cold temperatures freeze out most of the water vapor, the most important greenhouse gas. The quantity of water in the air during frigid glacial periods must be greatly reduced by this freeze-out. Water is such a strong greenhouse absorber that the decrease in water vapor would overshadow any radiation trapping effects from large changes in the other greenhouse gases.

Freezing most of the water out of the atmosphere would greatly decrease greenhouse heating. During ice ages, loss of greenhouse heating from water overwhelms any greenhouse changes due to carbon dioxide or methane. Changes in water vapor drive down ice age heating, and changes in carbon dioxide, such as the concentration found in ice cores, are lost in the larger water vapor effects.

This is an important observation. It tells us that decreases in the carbon dioxide and methane were not the driving force behind the cooling of the weather. Again, this does not indicate that increases in these gases will not cause the world to heat up, but it does suggest the need for further thought when pointing to ice age decreases in carbon dioxide as proof of greenhouse theories. The fact that two things vary together does not mean that one causes the other. The use of electric power correlates well with world population increases, but that does not mean electricity causes babies. Perhaps neither greenhouse gas concentrations nor temperature changes caused the other; some third factor could have caused both. Perhaps cold temperatures caused the decreased concentrations of carbon dioxide and methane shown in Figure 9, instead of vice-versa.

According to the figure, methane concentrations correlate better with temperatures than carbon dioxide. This is surprising, because at these concentration levels, carbon dioxide is a stronger driving force. It causes more greenhouse heating than methane. If the temperatures were driven by the concentrations of greenhouse gases, the best correlation should involve the more important greenhouse gas, carbon dioxide. Concentrations of deuterium, an isotope of hydrogen, correlate better than either carbon dioxide or methane. Yet no one suggests that the ice ages were caused by decreases in atmospheric concentrations of deuterium.

These variations in carbon dioxide and methane occurred long before humans were affecting global climates. It is reasonable to wonder what connection exists between worldwide temperatures and concentrations for these gases.

Methane comes from animal digestive systems, insects, and decaying plant material. During an ice age, the number of animals and insects decreases, just as flies and mosquitoes disappear in winter. Animal and insect production of methane should then decrease during an ice age. Also, fewer plants would grow, so decay processes would be retarded by the cold temperatures. This means lowered methane production. Once in the atmosphere, methane has a lifetime of just ten years, so production and atmospheric concentrations of methane would quickly come to equilibrium.

From these considerations it sounds like global temperatures drive the concentration of methane, rather than vice versa. Perhaps the same holds true for carbon dioxide.

As we've noted in an earlier chapter, the driving forces between carbon dioxide concentrations and temperatures are more difficult to trace. Scientists are currently unable to account for the whereabouts of half the excess carbon dioxide, which means that some aspects of the carbon dioxide cycle remain unknown. Still, consider what we do know. The ocean is the largest sink for carbon dioxide. People often worry about whether increased temperatures could decrease the amount of carbon dioxide that dissolves in the ocean. As the oceans become warmer, the solubility of carbon dioxide decreases.

If so, cold glacial waters might absorb more carbon dioxide, putting more carbon into the oceans and leaving less in the air. Or perhaps less carbon dioxide might be released from decaying matter in the soil. There is just too much uncertainty to understand what triggers changes in carbon dioxide concentrations during glacial periods. However, we can conclude that those who claim decreased carbon dioxide concentrations during ice ages are proof of the greenhouse gas connection may be wrong.

Increasing greenhouse gases can cause the world to warm; decreasing them can cause the world to cool. However, that does not justify the claim that the world cooled during ice ages because less greenhouse gases existed. The relationship between glaciers and gases remains a matter of speculation.

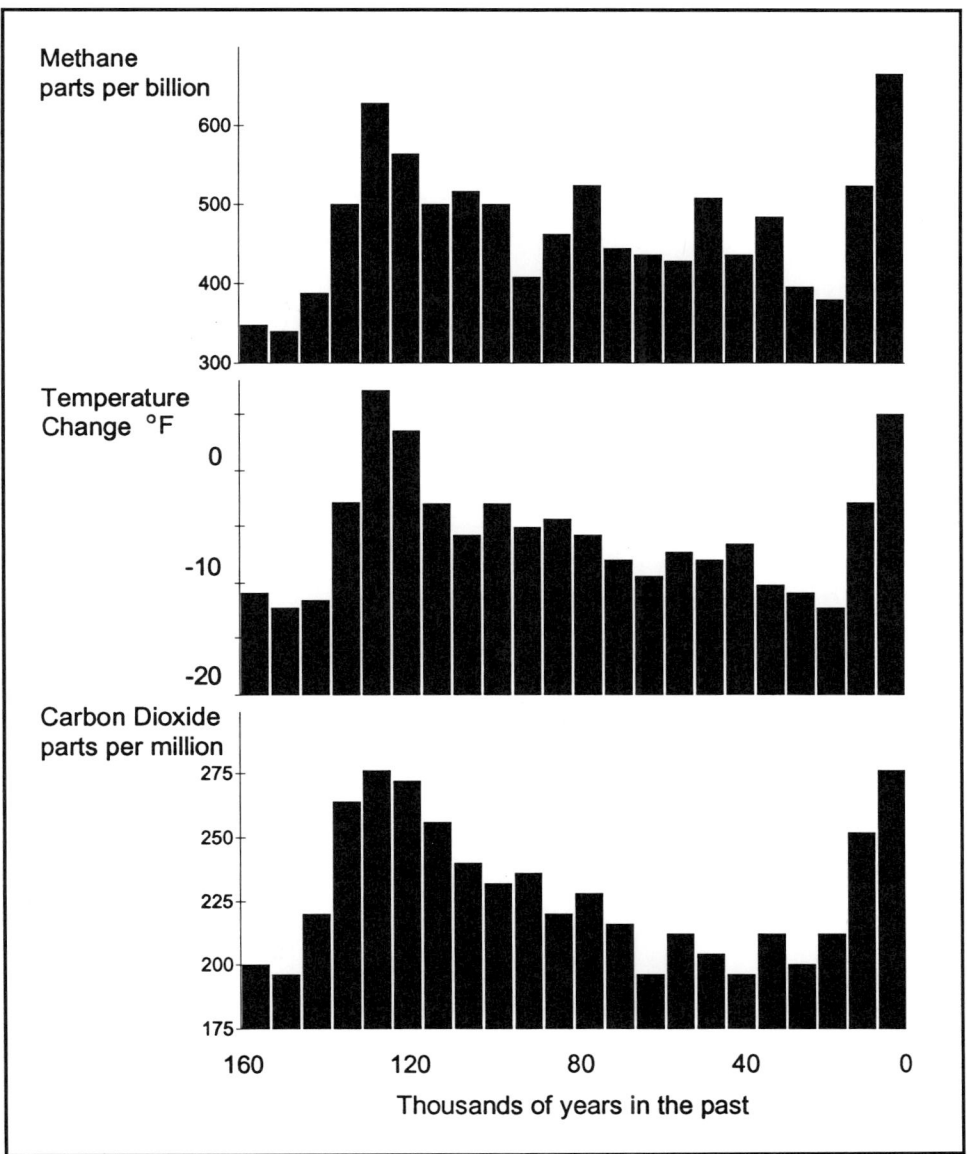

FIGURE 9 *Temperatures are often compared to concentrations of carbon dioxide and methane. The three do seem to go together, but this does not indicate whether temperatures cause gas concentrations to increase or whether increased gas concentrations cause temperatures to increase. The worrisome thing is that both carbon dioxide and methane are at higher levels now than ever before in history. The zero mark on the middle chart represents current temperature levels.*

Oceans and the Weather

New circulation patterns could ruin the tourist trade.

Oceans are seldom mentioned when people list the potential disasters we may cause through environmental abuse. When they are mentioned, it is usually to assert that oceans are big enough to absorb all the extra carbon dioxide we release. Some speakers invoke their size in hopeful rather than factual terms, forwarding oceans as a safeguard that will save us from our own follies. As sad and disappointing as it may be, oceans are vulnerable to change. Shared processes tie oceans and climate together, making both equally vulnerable.

We must recognize that the actions we take on land affect local climates. Farming practices that turn productive land into deserts exemplify local, human-driven climate change. It is important to remember that these local changes are going on throughout the world; their combined effect may eventually lead to a global change. Perhaps the oceans are postponing the day of change, but they are certainly not canceling it.

Oceans act like a big flywheel that dampens rapid global climate changes. The dampening effect is both a blessing and a curse. We're blessed because oceans delay the effect of human-made climatic change. We may be cursed because, once change starts, decades are needed in order to detect and reverse the changes.

In the days of Galileo, everyone believed that the earth occupied the center of the universe. Galileo's problems began when he denied that reassuring thought. He claimed that the earth rotates around the sun, and the fierce resistance to his rather obvious observation illustrates a weakness in the human character. We prefer to think of ourselves as the be-all and the end-all of creation—and we live on land. The fact that we dwell on land may prevent us us from recognizing the oceans' important role in determining the earth's climate.

Most people have the notion that oceans have little affect on climate. This amounts to vanity. We are humans who live on land and "control" our environment. Our ego dictates that land must be the center of the world, just as people once thought the earth was the center of the universe.

Most people also believe that oceans are so big and powerful that we have little impact upon them. So, on the one hand, we relegate the oceans to minor importance; on the other, we consider them too big to change. Ours is a myopic view of oceans.

Oceans cover 70 percent of our planet's surface. The ocean surfaces are dark—sometimes more black than blue—and they reflect a smaller fraction of solar heat than land. This means that oceans are the world's heat engine. Because they are so large, and because they absorb solar heat so effectively, they drive the earth's climate.

Oceans also provide the ultimate thermostat for controlling solar heating—clouds. If the surface of the ocean is heated, water evaporates. The water vapor forms clouds that block out part of the incoming solar radiation. This does not mean that oceans control all clouds, but oceans do tend to be more cloudy than inland areas.

Oceans magnify their climatic role by absorbing solar energy and carrying heat to other parts of the globe. In the big picture, oceans capture heat in the sun-rich tropics. Currents carry the heat to the polar regions, where the sun provides less heat. Those are the currents should worry about. Even a small change in ocean currents could mean a new ice age or a dearth of rainfall over entire continents.

Winds drive surface currents. If global wind patterns change, ocean surface currents change, too. Of course, other factors are involved, but winds are the primary driving force. Changing wind patterns by cutting rain forests or overgrazing land means we face the possibility that ocean currents will respond by altering their flow. No one knows just how great the risk is. In fact, if we acknowledge that much remains unknown about our weather, we must also admit that we know almost nothing about oceans. In all probability, this is because our egocentric nature provides us

with little incentive for understanding oceans. After all fish, not people, live in oceans. If not for submarine warfare, we might be even more ignorant about the oceans.

One particularly troubling notion is this: Ocean currents are like giant rivers. We expect weather and climate to change slowly, but we know that rivers can change their courses suddenly and dramatically. If we put a little too much stress on the ocean currents, they might switch into completely new patterns, just like the river. This might leave California in the cold, or cut off moisture from the Gulf of Mexico. We depend on the ocean-moderated climate of California for many of our fruits and vegetables. We depend on the Gulf moisture to grow the wheat and corn in our breadbasket states.

The El Nino cycles of drought and heat provide one example of how oceans can influence weather. When ocean temperatures off the western coast of South America increase to a few degrees above normal, an El Nino is created. Although the United States is far away, El Nino reduces rain in the corn belt, causes floods in the South, and spreads heat waves over much of the country. Other parts of the world suffer, too.

Oceans also grow living things and provide long-term chemical storage. On the average, oceans produce less plant and animal life per square mile than land, but the oceans are much larger, so on an overall basis, they produce more tons of plant and animal life each year than the land. We worry about cutting rain forests and releasing the carbon stored in the trees. Oceans store hundreds of times more carbon than the rain forests. Sea water stores carbon as several different compounds, but they all trace back to carbon dioxide. Oceans also place excess carbon into longer-term storage by producing giant formations of limestone and other rocks.

The ocean contains highly productive areas, but it also consists of large areas that are almost devoid of life—just as land consists of both garden areas and deserts. The non-productive areas are called ocean deserts. The desert-like regions contain have plenty of water, of course, and the sun provides the opportunity for plant growth. The problem here is a lack of nutrients. Almost total depletion of key nutrients makes large parts of the ocean about as productive as a desert. Without those nutrients, nothing grows.

People are not to blame for the depleted nutrients; it is the ocean itself that causes the problem. Microscopic plants like phytoplankton head the ocean food chain, because their photosynthesis converts sunlight to biological material. Photosynthesis demands light, but the sun only reaches the upper layers of the ocean, so plants can only grow near the surface. Plants provide the food animals must have in order to grow, but both plants and animals eventually die. After death they settle to the bottom, taking their minerals and nutrients along with them. This settling of dead matter and fecal material removes nutrients. Continued transfer of matter from the upper layers eventually depletes even the surface waters of the materials required for new plant growth. With no plants to feed the food chain, the animals wither away and ocean deserts appear.

Of course, not all ocean areas are deserts. In some regions, water rises from the bottom, bringing settled minerals and nutrients back to the surface. These up-welling regions of the ocean abound with life and provide rich fishing grounds. However, if surface currents change, the up-welling currents might disappear, too.

Even though winds drive ocean surface currents, the currents seldom move in the same direction as the wind. Besides surface currents, oceans consist of deeper circulations driven by other forces. Interactions among all these currents determine where the rich bottom-waters come to the surface and create ocean gardens.

The Antarctic is one area that blooms because of up-welling bottom-water that is rich in nutrients. In spite of the frigid, almost freezing, water temperature and reduced sunshine, part of the Antarctic Ocean teems with life, ranging from single-celled plants to whales. Heavy bottom-waters rise in these life-rich zones. The cold air above cools the surface waters, causing them to become even heavier and more dense than the bottom-water. Ice formation also plays its part, because freezing water accepts less than its full share of ocean salt.

The Antarctic ozone hole could have serious effects on this rich life-growth region. In the absence of ozone, solar ultraviolet can penetrate ecologically important depths of sea water. Phytoplankton absorb visible sunlight and provide the photosynthesis that heads the oceanic food cycle. If the ozone hole lets ultraviolet radiation through the atmosphere, that radiation could damage the phytoplankton, or at least decrease their photosynthetic production.

Phytoplankton cannot sense the UV, and the radiation could destroy some of those near the surface. Other varieties live in deeper waters and might survive, but less light penetrates to the deeper waters, meaning that less photosynthesis occurs. It seems likely that reduced photosynthetic production in the Antarctic region would reduce the productivity of the oceans, but maybe not.

Suppose the productivity of the Antarctic oceans decreased due to the ozone hole. Reduced productivity also means that nutrients would not be depleted as quickly. The water would flow away in surface currents and take its nutrients with it. The previously unavailable nutrients could then make other areas of the ocean productive. The basic idea is that if nothing eats nutrients in one spot, something else will eat those nutrients elsewhere. No one can say for certain that this would happen, but it appears that unused nutrients, which set the ultimate limit to ocean productivity, will eventually be used somewhere in the ocean. Lack of nutrients is the only reason that some ocean areas are non-productive. All the lost productivity in the Antarctic might just move to some other part of the ocean. The next question is whether the new area would make more efficient or less efficient use of the nutrients.

Scientists have always assumed that most of the carbon dioxide that disappears from the air must be absorbed by the oceans, casting oceans in the role of janitors that clean up our messes. Recently, some are questioning this assumption; and we often debate how long the oceans will continue to clean up after us. In fact, the topic gets more discussion in some publications than it deserves. Here is why.

Surface waters absorb carbon dioxide. After a few years, these surface waters cool and descend into the deep ocean. They do not emerge from the bottom for about a thousand years. When nutrient-rich bottom-water rises today and replaces surface waters, it is subject to modern atmospheric carbon dioxide concentrations, which are much greater than they were 1,000 years ago. The quantity of carbon dioxide already absorbed in these waters is unaffected by the increased carbon dioxide production of the last couple of hundred years. With the continuous replenishment of bottom-water, surface waters can probably continue absorbing as much carbon dioxide for the next few hundred years as they have for the last 200 years.

The main threat to continued absorption of carbon dioxide by surface waters is global warming itself. Warm water absorbs less carbon dioxide than cold water. If global warming takes place, then the oceans might absorb less of our excess carbon dioxide production, in turn causing an acceleration in global warming.

As an aside, much of the excess carbon dioxide we now put into the air simply stays there. Neither the oceans nor anything else absorb it. In fact, about half the carbon dioxide released since the Industrial Revolution is still in the atmosphere, causing the current increase in atmospheric concentration. Absorption by the ocean removes a little of the carbon dioxide, cutting down on the rate of increase, but only by a small amount.

Heat absorption by the oceans has probably kept alive the argument about whether greenhouse warming has arrived. The oceans absorb so much heat that air temperatures increase just a little—so little that no one can be sure they are not simply fluctuating upward for a while. Without significant climatic temperature increases, the assertion that greater concentrations of greenhouse gases will lead to hotter days for the earth seems a pretty weak statement, and so the argument continues.

Oceans can hold climatic temperatures down in this manner: If the air warms a little, ocean surface waters warm, too. The heat absorbed by the ocean waters prevents the air from warming further. Then these surface waters, along with the heat they absorbed from the air, descend into the ocean depths for 1,000 years. They take part of the excess carbon dioxide with them, but they also take part of the greenhouse heat trapped by the carbon dioxide. Without this flywheel to slow climate change, greenhouse heating would be larger and easier to verify.

Talking about surface currents and

other oceanographic topics may seem unrelated to the ways in which humans alter their climate. The point is that oceans and weather feed on each other, especially through ocean currents, and these currents exist in a delicate balance. Rivers are contained by banks. Ocean currents have no banks to give them direction. We cannot convincingly explain why currents go where they go. A "little" change in climate might cause major shifts in oceanic flow patterns. These major shifts might create other major changes, like another ice age. Many experts believe that ice ages come about because of shifts in ocean currents.

In a way, oceanography has been the leftfield of science. Oceanographers have gone to sea to study the biology of the ocean and its physical condition, while other scientific disciplines rarely paid any attention. But global concerns are bringing oceanography into the mainstream of geophysics, and many researchers are realizing just how badly they have neglected oceanography.

OCEAN SURFACE CURRENTS

Woods Hole Oceanographic Institute has dropped tens of thousands of bottles into the oceans, just to see where the currents carried them. This may sound like scientific littering, but each bottle contains a note offering a reward. On average, someone claims the reward for about ten percent of the bottles within three years. Other reward requests may go back 10 or 20 years. People who send in the reward form receive 50 cents. It costs almost that much to mail the form! Presumably, someone at Woods Hole collects these forms and fills notebooks with data indicating when and where the bottles were found. This may seem like a strange way to conduct science and spend taxpayers' money, but the person doing it undoubtedly thinks this work is vital to the world. And he or she may be right. The man who filled notebooks with data on ozone in the stratosphere seemed to be wasting money until the ozone hole appeared.

Literature includes a number of stories about putting one's name in bottle, casting it into the sea, then waiting for a reply from some faraway, romantic place. Such stories have some basis in fact. Ocean currents carry out very complex circulations, and given enough time a reply could come from almost anywhere. "Anywhere" includes a spot ten miles down the beach, because the currents eventually return to their starting point after circling the globe. Imagine waiting five years, then getting a reply from ten miles away—or finding the bottle yourself just where you tossed it in. You would probably be unaware that the message had circulated through all seven seas and come back.

Although surface currents return to their starting point, the ocean water itself may not come back. As the currents circulate, the water heats, cools, evaporates, and sometimes gets more salty. During these processes, part of the water cools and gets dense enough to sink to lower levels in the ocean, where circulations are different. The bottle, however, floats on. Convergence and divergence zones also affect the water's movement. Convergence zones are locations at which currents come together from different directions. Since all the incoming water has to go somewhere, some of it to sinks to lower levels. After the excess water gets to lower levels, it flows away from the convergence zones.

Divergence zones are more important. In these areas, there is more surface water moving out than coming in. Divergence zones occur when water rises from below, as from a spring. Surface currents flow away in several directions, carrying the rising water. Within divergence zones, rising water brings up the nutrients needed to feed life-rich growth zones.

As we've stated, surface currents get their drive from surface winds, but wind does not determine the direction of flow. Phenomena called Coriolis forces cause flows to deviate to the right of the winds in the northern hemisphere, and to the left in the southern hemisphere. Because of these deviations, surface current patterns differ greatly from surface wind patterns.

Coriolis forces come about because the world rotates. Because of rotation, things appear to move in curves instead of straight lines. Throwing newspapers from a moving car provides a good example of how Coriolis forces are created. If you think back to throwing newspapers, you may recall how difficult it was to make the paper fall onto the porch. After the newspaper was thrown, it seemed to move in a curve, not along the expected straight line. Of course, the person standing on the porch saw the paper move in a straight line. To you, however, the paper seemed to sail along a curve, because you were watching from a moving platform—the car.

Coriolis forces make both winds and ocean currents appear to deviate from a straight line because our platform—the earth—is moving. How much of a curve is evident depends on how fast the wind or ocean current is moving. It also depends on how far the wind or current is from the equator—the apparent rotation of the world is different at the poles than on the equator.

Because winds drive surface currents, a look at winds can help us understand how Coriolis forces work. Figure 1A shows the average global wind pattern at the surface. Notice the spiral patterns, such as the one off the western coast of the United States or the pair of spirals in the Atlantic off the northwestern and southwestern coasts of Africa. Both winds and currents move in spirals, or in paths that look something like spirals. Coriolis forces affect both.

These spirals occur because the earth has areas of consistently high and low atmospheric pressure, even at sea level. The rule, often asserted in high school science classes, is that atmospheric pressure at sea level is the same everywhere, at least when averaged over time. The instructors neglect to mention that some regions refuse to follow this rule. These areas may be at sea level, but some always seem to have too much or too little pressure. Figure 1B indicates the average atmospheric pressure at sea level. The averages extend over many years, and show that some regions consistently have high or low atmospheric pressure. The curves in this figure are called Isobar curves. They are lines along which the pressure is constant. Lines are labeled in millibars deviation from the standard atmospheric pressure of 1013 millibars, which is what sea level pressure is supposed to be. Notice that the high and low pressure regions are centered over areas in which wind spirals appear. This is not by chance.

You might expect wind to flow directly away from high pressure regions instead of spiraling around them. Air tries to escape high pressure zones and move to lower pressure regions. Winds try to move air in a direction away from the high pressure, but Coriolis forces cause the winds to deviate. They force the wind to move to the right of the direction in which the wind was headed. The deviation makes the wind move in a new direction, but the Coriolis forces make it keep moving further to the right of its current direction. This continuous forcing to the right finally rotates the wind in a complete circle of sorts—really a spiral. In the southern hemisphere, deviations are to the left instead of to the right. The same Coriolis forces cause the circular wind patterns we see in tornadoes and hurricanes. They also cause spiral patterns in

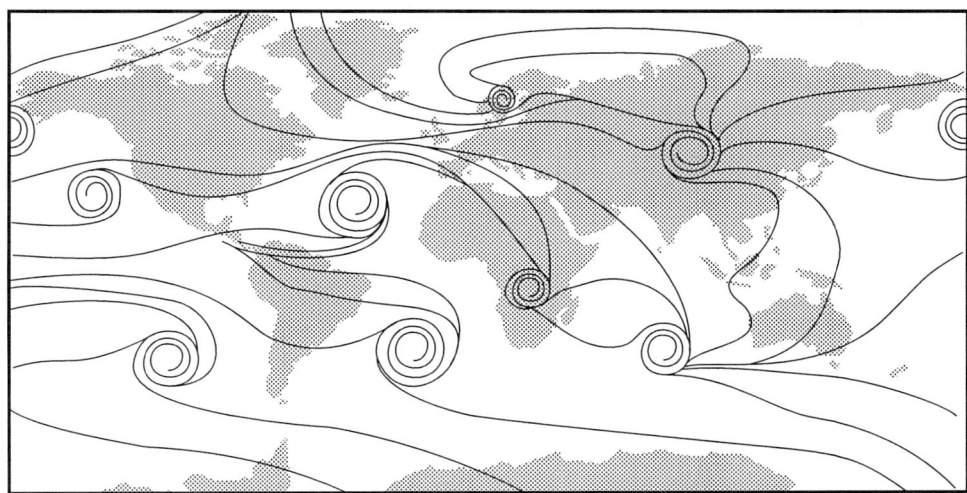

FIGURE 1A *The lines show the average air flow throughout the world. Winds circulate around the high and low pressure regions shown in the previous figures. The direction of circulation around highs and lows reverses in the southern hemisphere.*

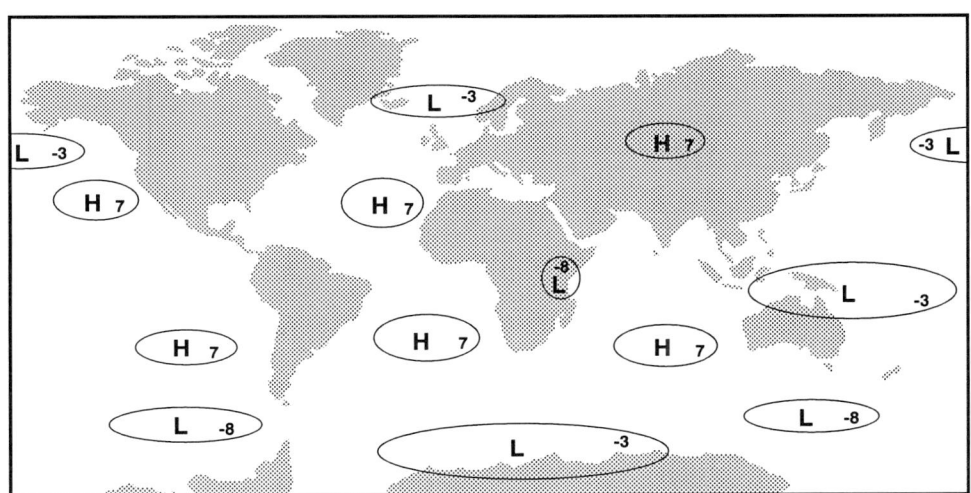

FIGURE 1B *Sea level barometric pressure is not the same everywhere. Average pressures are high at some locations and low at others. H and L indicate consistently high and low pressure regions. The numbers show how many millibars high or low these areas are. Normal sea level pressure is 1013 millibars. The averages were taken over many years.*

ocean surface currents, but ocean spiral patterns are given another name. Oceanographers call them gyres.

Strange things happen in the world of flows. The way in which wind spirals around high and low pressure regions is one example. Air refuses to flow out of high pressure and directly toward the low pressure regions, as a reasonable person might expect. As another example, consider the flow of the Mississippi River.

The world is a little fat around the equator—some might say pot-bellied. It is not the perfect sphere implied when teachers say the world is round. The Mississippi flows south, and in the process gets closer to the equator. Earth's pot belly ensures that the nearer the river gets to the equator, the farther its surface is from the interior center of the earth. In a sense, then, the Mississippi appears to flow uphill. If measured in terms of distance from the center of the earth, the Mississippi *does* flow uphill. It can do so because the centripetal force of the earth's rotation affects the force of gravity. It makes gravity a little less powerful near the equator, and this may explain the earth's bulging middle.

Here is another strange actuality. The ocean has high and low spots. For example, the high pressure regions of the atmosphere press down on the ocean surface, depressing it in regions of high atmospheric surface pressure. The high pressure region off the western coast of the United States depresses the sea surface by almost two feet. Low pressure regions, on the other hand, "suck up" the ocean surface by corresponding amounts.

A foot or two variation in the height of the ocean may not seem significant, given the size of the oceans. However, what implications does this have for our concept of sea level? If the surface of the ocean varies due to atmospheric pressure, how can anyone be sure that sea level is changing? Sea level fluctuations at any given location could be due to changes in atmospheric pressure patterns.

Current estimates indicate that global warming might cause sea level to increase by three feet. Notice that local increases of almost this magnitude occur as a result of atmospheric pressure distributions and the flow patterns of ocean currents.

Despite the slightly flawed argument that the Mississippi River flows uphill, we all expect water to flow downhill. The next question is this: How do Coriolis forces produce flow in the ocean? The answer is

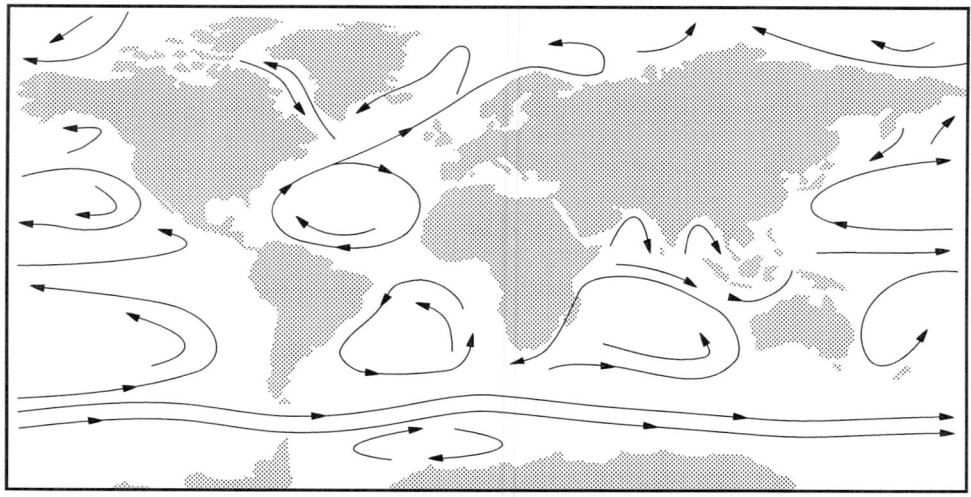

FIGURE 2 *Winds drive the ocean surface currents in the patterns shown. Although they are driven by the winds, the ocean patterns differ from wind patterns. Wind can blow across ocean or land, but continents block ocean currents. This results in the almost circular flowing gyres in the Atlantic, Pacific, and Indian oceans.*

that the forces cause water to pile up, making the sea a little higher in some places. Ocean currents can then flow downhill from these pileups (in combination with wind-driven flow).

Radar altimeters in satellites have measured the slope of the sea surface across the Gulf Stream. At times, these measurements indicate that ocean surfaces slope by three feet over 70 miles. The slope causes surface water to flow at speeds of three feet per second at latitudes of 43°N. Such flows are called geostrophic currents.

Figure 2 shows the general scheme of surface currents. A giant gyre dominates each ocean (except the Arctic), flowing clockwise in the northern hemisphere and counterclockwise in the southern hemisphere. These surface currents would be unimportant were it not for the temperature variation in the ocean and the way the currents redistribute heat around the globe. The redistribution of heat makes these currents vital to our climate, especially along the coasts.

Figure 3 shows worldwide average surface temperatures in January, which is winter in the northern hemisphere. Warm water flowing from the Gulf of Mexico warms temperatures along the eastern coast of the United States, then spreads across the Atlantic. The warm Gulf Stream ensures that temperatures in Scotland and Norway are comparable to temperatures in Maryland and Virginia. Contrast the temperature in Sweden with the much colder temperatures in Siberia. Both lie at comparable northern latitudes.

It may take several years for water to complete a circulation around a gyre. Estimates say it takes five years to make the full circuit around the Gulf Stream. The speeds of surface currents vary, but they typically range from three to six feet per second.

The most confusing thing about the surface current off the California coast is the number of names applied to it. It is variously called the Kuroshio, Japanese, and California current. Whatever the name, it

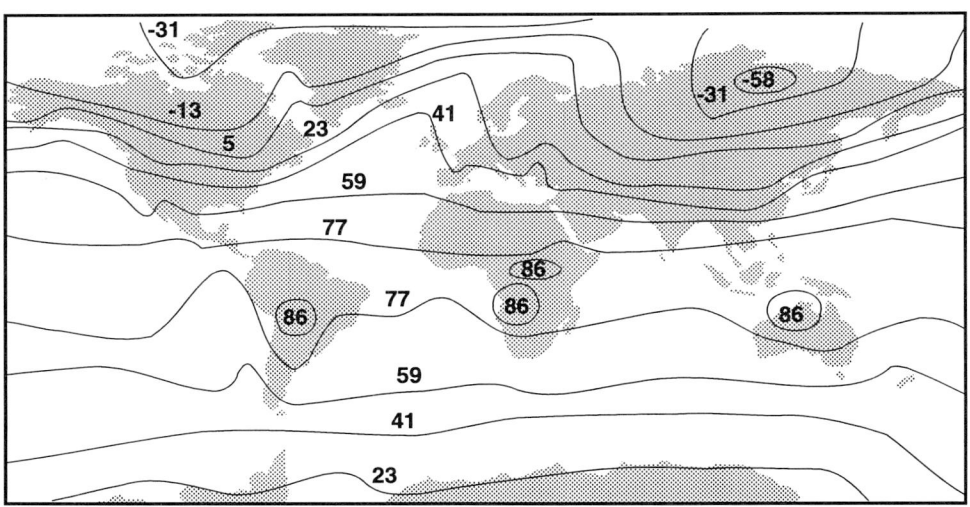

FIGURE 3 *The lines show the average surface temperatures in January. Notice that the warm Gulf current makes the North Atlantic and Scandinavia much warmer than places like Siberia, which are equally far north. The cold Antarctic current causes colder temperatures off the west coast of South America.*

is a warm current containing water heated while flowing from east to west in the tropical Pacific. When the current bumps into Asia, it turns north, flows past Japan, back across the Pacific, and down along California. In spite of its long trip, the water is still warm enough to cause warm winters in California.

Not all currents are warm. The Peru Current off the western coast of South America is a cold current. Water cooled during its trip along the edges of the Antarctic flows along the coast. The current causes the northward bulge in colder temperatures seen in Figure 3 along the coast of Chile and Peru.

Without surface currents, the polar regions would be even colder, and the tropics would be even hotter. Waters warmed by tropical sunshine carry heat to the polar and temperate zones of the world. The transfer of that heat to the air involves more than water heating the air above it. Much direct warming of the air occurs, but most of the transfer takes place through evaporation and rain.

When water evaporates from the surface of the ocean, energy loss from evaporation cools the sea water. To vaporize a pound of water requires 971 BTUs of energy. Stated another way, each pound of water evaporated from the ocean cools 971 pounds of the remaining water by a full degree Fahrenheit. Water vapor that escapes through evaporation eventually condenses and falls as rain. Conservation of energy laws dictate that the condensation of a pound of water must release the same amount of energy that was used to evaporate it. Thus, water vapor extracts heat from the ocean when it evaporates, then releases the same amount of heat into the atmosphere when it condenses.

Suppose that water evaporates, sends its vapor straight up, condenses, then falls back into the ocean as rain. Conditions would not be as they were prior to the evaporation. Ocean waters would be cooler, because the evaporation process cooled them. Air would be warmer, because it was heated by the condensing water vapor. The heat warming the air is the same amount of heat the water vapor took from the water when it evaporated. The water is now back in the ocean where it started. The only changes are a net cooling of the ocean and a net heating of the air.

In the real world, rain seldom falls straight back into the ocean. Rain could fall over land or far downwind of the spot where the water vapor evaporated. The water vapor would still warm the air when it condensed, but this might take place far from the evaporation site. Ultimately, the ocean transfers more heat through evaporation than through direct heating of the air above.

EL NINO AND OTHER OCEANIC EFFECTS

Unusual weather patterns sometimes correlate with conditions at the sea surface. Conversely, when some particular condition develops at sea, weather records indicate that something generally happens to the weather. The weather change may involve temperature, winds, or rainfall. The change may be nearby or far away. It may cover a small area or an entire region.

Statistics are the only way to prove that a relation exists between the weather and an ocean-based event. Once they recognize such a relationship, meteorologists usually conclude that one event causes the other. However, their rationale is normally too weak to stand on its own without statistical proof.

I mention these tie-ins to emphasize the fact that changes in climate and ocean circulations can interact and feed upon each other. Weather events typically last a few days and disappear. Ocean current and surface events last much longer, and noting this difference is the key to recognizing tie-ins. Unusual weather that lasts too long is probably produced by a tie-in with something else.

Ocean tie-ins are worth checking out. The process may be difficult, though, because the two events are sometimes far apart—as when warmer water off Peru causes droughts in Kansas. There is one other aspect to these tie-ins—both events may occur simultaneously, so it may not be sensible to assert that the ocean causes the weather or vice versa. The two feed on each other. Who can say which is the mother and which the daughter?

When one event starts first and the other comes later, distinguishing the mother from the daughter is easier, but statistics still help. Statistical analysis of events associated with sea weather suggests that atmospheric anomalies often induce anomalies in the ocean. This suggests that the atmosphere is the initiating force, or cause. Ocean patterns lock in to the weather pattern, then generate forces that drive anomalous weather patterns.

The El Nino effect is such a case, and it presents a fascinating story. It is fascinating because it was talked about for a long time, because its its aspects were identified slowly and with great difficulty, and because it ties ocean phenomena with remote weather events. The effects of El Nino in Peru were documented during the Spanish conquest. The name "El Nino," applied to a warm, southward ocean current, goes back to the 19th century.

The El Nino current replaces the cool, northward-flowing Peru current from time to time. In the eyes of fishermen from an earlier time, the occurrence had something to do with the Christ Child, because the warm current frequently occurred near Christmas. Thus, it was given the Spanish name for the Christ Child.

The first recognized tie-in with weather came near the turn of the century. Peruvian geographers noticed that, in those years when the current was particularly strong, heavy rains fell in the nearby desert. Soon after that, Peruvians started exploiting the guano of marine birds for fertilizer and realized that bird populations declined during times of strong El Nino-triggered events. The bird populations declined because there were fewer anchovies and sardines for birds to eat during these events.

Industry in Peru expanded up the food chain. From digging guano generated by sea birds that fed on anchovies off the coast, the Peruvians added catching the anchovies to their commerce. With this double exploitation in place, many people began to recognize the connection between the El Nino current and anchovy production. In the early days, there were so many anchovies that fishermen caught them to produce fish meal—ground up, dehydrated fish used as an additive in livestock feed. By 1971, the Peruvian fishing fleet was a giant industry. In that year, Peru caught 13 million tons of anchovies. Then came the El Nino of 1972-73, and production never recovered. A weather-related event and overfishing combined to cripple an industry.

While knowledge of El Nino currents was common in Peru, related events were taking place across the Pacific. At the turn of the century, a British climatologist named Walker discovered what he called the southern oscillation, "a swaying of mass between Australia and the eastern Pacific." In years when atmospheric pressures were high in the Australia-Indonesia area, rainfall was low. In years when high pressures shifted from Australia-Indonesia to the eastern Pacific, rainfalls were heavy. Walker later expanded his studies and tied in this southern oscillation with other climatic anomalies from around the globe.

However, he failed to note the connection with the El Nino current.

In the 1950s, someone finally noticed a connection between the southern oscillation and ocean temperatures. It is easy to understand why it took so long to tie atmospheric pressure in the eastern Pacific to water temperatures off the coast of Peru. The two areas are separated by thousands of miles of ocean, and there is a delay between the southern oscillation and the El Nino current. The marvel is that anyone tied them together at all.

Here is how the connection works. For a couple of years during the height of the southern oscillation, brisk western trade winds blow along the equator in the Pacific. The winds, which occur a year or so before the El Nino, build up a thicker-than-normal layer of warm surface water in the western Pacific. Cool, upwelling water covers the eastern Pacific. It is this upwelling water, filled with nutrients, that grows such large populations of anchovies.

Everything is fine until tropical storms near Indonesia migrate eastward, inducing an eastward ocean current. The storm-induced current carries the layer of warm water from the western Pacific to the eastern Pacific. Changes caused by the warm current cause the storm centers to follow along. The storms migrate even farther east, producing more warming of the ocean surface. Weather events and surface warming feed on each other for several months until the entire Pacific equatorial zone is warmed. Sea levels in the eastern Pacific increase by six inches to a foot, and the sea surface temperatures increase by as much as 14°F. These conditions continue for a year or two before things return to normal. El Nino events occur every two to seven years, but vary greatly in intensity.

The world had paid little attention to El Nino until the bad weather of 1982-83. Drought in Australia and floods in western South America were only part of the weather crisis. Europe suffered from record summer temperatures, and the American midwest was both dry and hot. Summer weather cost midwestern farmers greatly by reducing soybean and corn production. Very cold late-fall temperatures occurred throughout the United States. Since then, we have heard many forecasts for El Nino events, but the forecasts are not yet very accurate.

El Nino is the best known oceanic-atmospheric weather event, but there are many others. Sea surface temperatures, for example, provide one critical parameter for predicting the formation of hurricanes. Increased sea surface temperatures are also blamed for the terrible drought in the African Sahel. There, the warmer ocean surfaces in the South Atlantic, the Indian, and the Southeast Pacific oceans combined with cooler-than-normal temperatures in the North Atlantic and the Pacific to change weather patterns.

Ocean surface temperatures south of Newfoundland seem to control the direction of air circulation over central Europe. The European circulations seem to reverse directions as temperature anomalies south of Newfoundland change from warmer than usual to cooler than usual. The temperature anomaly may cause the reversals by its effect on the position of the jet stream.

For some unknown reason, a seasonal relationship exists between the summer ocean temperature in the Gulf of Alaska and the circulation of air over the United States during the following fall and winter. Such a long delay between the event and its effects makes this relation interesting. At least in this case, we know the ocean temperature is affecting the weather. Of course, in a few years, someone may find a weather event in some other part of the world that determines summer ocean temperatures in the Gulf of Alaska.

The point of this discussion of tie-ins is that weather and oceans interact, often in destructive ways that are far from obvious. We must therefore act cautiously to ensure that small local changes in weather do not upset the ocean patterns that decide weather in wider portions of the world. Acting cautiously is difficult, however, because we are only beginning to understand the connections between our actions, the weather, and the oceans.

DEEP OCEAN CIRCULATIONS

Surface currents in the oceans extend to depths of perhaps 600 feet. Below these depths, much slower currents of sea water flow in completely different patterns. The deep currents are sometimes called thermohaline circulations. Thermohaline is a composite word meaning "thermal" and "salt." The cold temperatures and high salt content of deep, dense waters cause them to crawl slowly along the bottom of the sea. Surface currents involve only a small fraction of the water in the oceans—most ocean waters are found in the depths.

If we could spread the ocean's waters uniformly over the earth, we would create a world covered in its entirety with 8,800 feet of water. If we leave the land uncovered but make the depth of the ocean the same everywhere, the oceans would be 12,500 feet deep. The 600 feet or so of surface waters, then, makes up only a tiny fraction of the total ocean water.

The primary characteristic of deep ocean waters is monotony. Things happen slowly and without variation in the depths, but important things do happen.

Bottom-water, another name for deep ocean water, spends 1,000 years submerged well below surface currents before coming back to the surface. No glimmer of sunlight ever reaches it during this time. There is no day, no night, no seasons.

Surface water is at near freezing temperatures when it descends into the deep, and it is at about the same temperature 1,000 years later. It's a rather dull millennium.

Something does change while the water is submerged. The descending water contains almost no nutrients. Water coming up from the bottom is nutrient-rich, laden with the minerals that rain down on the bottom water for 1,000 years.

Some exotic creatures live in the deep water. Because they are in perpetual darkness, almost half the life forms provide their own light. These lights are great for luring prey, frightening enemies, and enabling the sexes to find each other. The other half of the deep-sea dwellers are blind. All have small bones, because bone formation requires vitamin D, which forms and functions where ultraviolet light is present.

Deep water becomes deep water when it reaches a greater density than the water beneath it, which causes it to sink. There are two ways in which the density of sea water increases. Water can increase its salt content, or it can undergo thermal contraction to become denser. When water cools, it contracts. The idea of contraction is a point of confusion to some, because we are told that pipes are busted when the water within them freezes and expands. That is true. Pure water contracts until a few degrees above freezing—then it starts to expand.

Sea water contains about 35 parts of salt per 1,000 parts of water. The salt content makes it more dense than fresh water, changes the temperature at which it freezes, and reduces the temperature at which additional cooling produces expansion instead of contraction. In short, sea water is different from fresh water. One good way to cool ocean water and make it dense enough to sink to the bottom is to send that water to the Arctic or Antarctic.

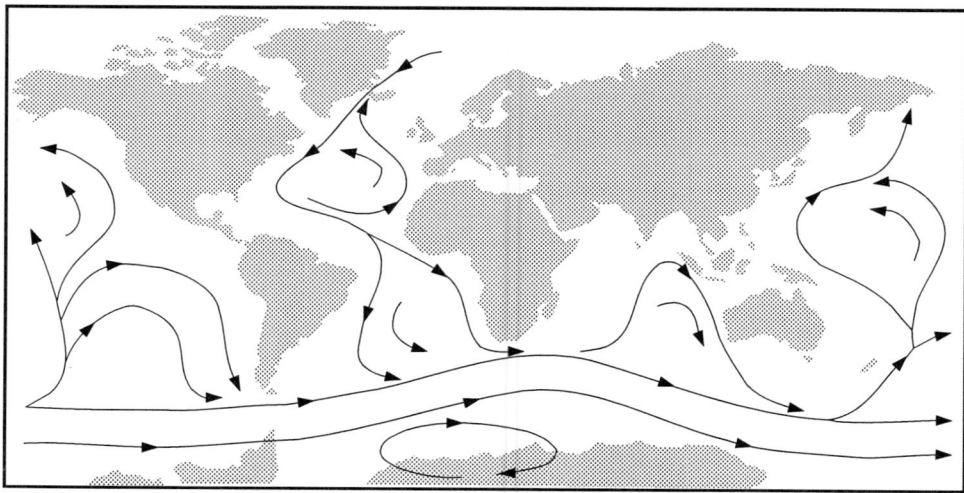

FIGURE 4 *Differences in temperature and salt concentration drive the circulation in the deep oceans. Cold, salty water descends off Greenland. It then circulates through the Atlantic, across to the Indian Ocean, and into the Pacific. The circulation is too slow to measure, so we have to estimate.*

There are two means of increasing the concentration of salt in ocean water—evaporation and ice formation. Both involve the removal of water while the salt remains behind. When water evaporates from the ocean surface, the surface waters that remain contain more salt. If the ocean were well-mixed, the increase in saltiness would be insignificant, but surface layers are mixed only to depths of about 100 feet. Just how deep the mixing goes depends on the violence of the wind and how much heat the surface gains or loses. You may recall that ocean currents carry solar heat absorbed in the tropics to the north, redistributing the heat largely through evaporation and rain. As a result, the oceans tend to become cooler and saltier as they move northward.

In some regions, evaporation can greatly increase the saltiness of the ocean. In the Red Sea, for example, very hot, dry, desert air flows across the narrow sea. Rapid evaporation and lack of circulation from the open ocean makes the Red Sea quite salty; the isolation of the Red Sea and its warm water temperatures prevent its waters from becoming part of the main bottom-water circulation. Evaporation in the Persian Gulf also produces water of high salinity.

The Mediterranean loses more water to evaporation than rivers can supply. Land encloses the Mediterranean except at the Strait of Gibraltar. The Mediterranean imports ocean waters through surface flow into the Strait, while exporting water that is high in salt content through the bottom of the Strait. Flows through the bottom of the Strait may be as great as six feet per second. This salty water spreads into the Atlantic at intermediate depths, which seems to explain why the Atlantic is more salty than the Pacific—even though only five percent of the Atlantic deep water comes from the Mediterranean.

Figure 4 shows flow patterns for the deep ocean water, but these patterns should be taken with a grain of salt, so to speak. Flow rates in the figure are estimates based on available measurements, model computations, and expert guesses. Very slow flow rates characterize the deep water. For the most part, the flows are too slow to measure. Notice that three regions are shown as source areas that produce bottom-water—two off of Greenland and one in the Antarctic's Weddel Sea.

The ocean currents are not steady, slow-moving flows. Sonic measurements of ocean water motion show that movements within the ocean behave much like our weather. There are underwater fronts, convective activity, and turbulent mixing of waters. Knowing that these events take place in water affords us a glimpse into the difficulty of trying to model the oceans. The models need many grids, a short time step, and complex interactions. Lack of computer power limits how well we can model the atmosphere, and the same shortage of power limits modeling of the ocean. Eventually the models of atmosphere and ocean must merge into a single model, because it is the ocean-atmosphere combination that produces weather.

The freezing of sea ice is critical to production of bottom-water. When sea water freezes, it does not freeze as a unit. Crystals of pure water freeze first, leaving their salt behind in the unfrozen water. As freezing proceeds, concentrations of salt-free crystals increase until they form a mass that encloses pockets of unfrozen brine water. As the freezing proceeds, it excludes most of the brine water. That brine in the unfrozen ocean water leaves the ocean water quite salty. The water was already cold, and now it is more salty and dense than the underlying water, so it sinks.

The reverse of an old saying applies here: Everything that goes down must come up. But deep waters do not re-emerge at the same location. Where these

waters come up has to do with the convergence of flows within the deep water. Convergence of surface currents, on the other hand, plays a role in deciding where water descends. When flows converge in a region, some water there must either rise or sink. Otherwise, the incoming water would simply pile up—which doesn't work too well with unfrozen water.

Surface flows seem to indicate that surface waters remain mostly within a single ocean, within their own gyres. Bottom flow does not work that way. Its flow is driven by density instead of winds, and the Arctic produces the densest water. This water finds its way to and through the South Atlantic. A couple of hundred years later, it reaches the Indian Ocean, moves on, and finally becomes part of the Pacific Ocean. After about 1,000 years, its trip is over. It resurfaces, bringing nutrients with it.

Arctic water is denser than Antarctic water because of the Antarctic continent. Ocean water flows under the Arctic ice. Some water freezes and leaves salt behind, which makes the remaining water more dense. The Antarctic continent covers the southern polar region and allows less water to freeze or thaw each year. Icebergs from the Antarctic region are different than Arctic icebergs. Those from the south are flat bars, instead of the oddly-shaped chunks of ice that come from Greenland and the north. The North Atlantic produces some 20 million cubic yards of deep water each second. The Weddel Sea produces even more bottom-water—about 40 million cubic yards per second.

What is the importance of bottom-water to global climate change? You may be relieved to learn that it is not of great significance. When life vanishes from the earth, the deep ocean will be the last place to become lifeless. Life may be monotonous in the depths, but it is stable. If life ceased on the top side, the bottom would not be affected for hundreds of years.

We may not care what goes into the deep, and what it does there may be completely uninteresting to humanity, but what comes up is important and vital—not only what comes up, but also *where* it comes up. The upwelling bottom-waters provide the nutrients that make parts of the ocean gardens, instead of deserts. If we cause surface currents to shift, the points of divergence that dictate where bottom-water comes up would also shift. This could destroy all of our rich fishing grounds. Generations would pass before new ones developed in other locations.

LIFE IN THE OCEANS

The first rule of farming is this: Put water on it, and it will grow. But that rule seems out of place in the ocean. Oceans always contain plenty of water, but there are large areas that "grow" very little. A farmer would tell you that these areas need fertilizer, and in essence that is correct. Water alone is not enough to grow plants, and animals depend on a food chain that eventually reaches back to plants. In the ocean, the food chain usually begins with some form of plankton.

Consider an isolated section of ocean. At first, everything is growing well. As time passes, some plants and animals die and their remains drift downward, away from the upper regions illuminated by sunlight. Even if nothing dies, fecal material will drift downward and carry away part of the minerals from the upper layer. It is only a matter of time until lack of minerals in the surface waters prevents growth there. That "mineral deficiency" exists in 97 percent of the oceans.

Two areas produce almost all of the ocean's organic matter—the continental shelves and the regions of upwelling bottom-water. The continental shelves are productive because rivers feed them with new minerals and nutrients. Water from the relatively shallow ocean bottom also mixes upward from the shelf, bringing nutrients to the surface.

As discussed earlier, regions of upwelling bottom-water get their minerals and nutrients from the rising water. Bottom-waters caught these materials as they settled from the upper regions of the ocean. Very few of the substances that settle toward the bottom reach the ocean floor as solid material. Chemical changes take place, and most of the material dissolves. Although these chemical changes require oxygen, it seems that there is always enough oxygen present to dissolve the droppings.

Other regions of the ocean are not completely without life, because they do receive a few nutrients from mixing continental shelf water, or from the annual turnover of the seasonal thermocline. Waves keep the topmost layer of the ocean well-mixed and at a uniform temperature. Just below this thin, well-mixed layer lies another layer, the thermocline, where temperature decreases with depth. Sunlight heats the surface layer, and some of the heat is mixed below the top layer. This produces the stable region where temperatures get colder at greater depths. When winter comes, the surface cools, and the seasonal thermocline breaks up. Its water then mixes with the surface layer.

The lower-level water from the thermocline contains some nutrients and minerals. The breakup enables the ocean surface to blossom for a brief period each spring—a burst of activity that consumes most of the available nutrients. Except for the spring flowering, most parts of the ocean remain almost dormant during the rest of the year. During the off-season, the standing crop of zooplankton may be just a third or half its maximum value.

"Productivity" as referred to here is the

mass of biological material that is produced. The ocean produces life containing about 100 billion tons of carbon each year—probably five times as much as the land produces. Of this 100 billion tons, only about 60 million tons consist of fish and shellfish. The rest involves much smaller life forms, like diatoms, plankton, and zooplankton.

Antarctic regions are particularly rich in life because the water there is so cold. At cold temperatures, more organisms can live on the same amount of food than is the case at higher temperatures. As a result, there are ten times more zooplankton in Antarctic waters than in tropical Atlantic waters.

Although unharvested by humans, such life in the oceans is critical to our weather. If life in the oceans should cease, atmospheric carbon dioxide concentrations would shoot up, because living matter provides a biological pump for removing carbon dioxide from the air and storing it in deep ocean waters.

The biological pump works as follows: Living matter in the upper layers of the ocean absorbs carbon dioxide. The absorption process maintains the carbon dioxide concentration at a level below saturation, which allows surface waters to continue to absorb carbon dioxide from the air. Living matter in these layers absorbs the carbon in the water and converts it into living material, thus removing the carbon from the water. Each living plant or creature eventually dies and sinks into the deep water, taking the carbon in its body along with it. Carrying the carbon downward means that the upper ocean contains lower concentrations of carbon compounds than the deeper ocean. If the upper waters contained as much carbon as the lower layers, the layers in contact with the air would absorb little or no carbon dioxide from the air.

Our assignment then, should we choose to accept it, consists of two parts: We should avoid poisoning life in the sea, because it provides our biological pump; and we must avoid upsetting ocean circulation patterns, because they are part of the control mechanism for our weather. Perhaps a third part of the assignment for some of us involves learning more about oceans, making the first two parts easier for the rest of us to live by.

▲ 11 ▼

Volcanoes

Meet the really big polluters.

There is something heroic about a volcano, about the way it just stands there, acting the way it wants to act without regard for humans. Volcanoes do things that are regarded as terrible when people do them. Volcanoes kill thousands for no reason at all; they pollute the atmosphere; and they callously accumulate troves of riches, including diamonds and gold fields.

Mercury vaporizes from the heat in rocks near volcanoes, sending mercury levels higher than the Occupational Safety and Health Administration would tolerate. Emissions of particulates, smoke, and sulfates are enough to make an Environmental Protection Agency inspector expire on the spot. And the way volcanoes can explode with more force than the biggest nuclear weapons—wiping out entire cities—must make military leaders envious. A volcano does all these things, yet people don't try to make it stop. They don't try to build one to use for their own devious purposes, and they don't even complain that volcanoes are a threat to our safety and environment.

Volcanoes are also the source of much beauty. Mt. Hood, Mt. Fuji, and Mt. Kilimanjaro are all peaks created by volcanoes. The tropical beauty of the entire Hawaiian island string emerged from the sea due to volcanic eruptions, as did the rugged landscape of Iceland. Rich agricultural lands often develop in volcanic soil. One study found that the volcanic areas of Indonesia are more thickly populated than the non-volcanic areas.

Stories about the power of volcanoes are very much a part of life and history. Mt. Vesuvius so completely covered Pompeii with volcanic ash that the city remained undiscovered and unreclaimed for 1,500 years.

Krakotoa, situated between the Indonesian islands of Java and Sumatra, exploded in 1883 with a force equivalent to 5,000 megatons of TNT. This compares to the 50-megaton yield of the largest nuclear weapon ever exploded.

Mt. Pelee in Martinique killed 30,000 people in 1902. The entire population of St. Pierre, except two, perished in the rush of hot air that swept over the town—air hot enough to set ships in the harbor on fire.

When Mt. St. Helens erupted, it blew 1,200 feet off the top of the mountain, leveled giant Douglas fir forests ten miles away, and created world-class landslides.

Another Indonesian volcano, Tambora, erupted in 1816, shooting massive amounts of dust into the stratosphere. The dust blocked enough sunlight to reduce temperatures as far away as Europe.

Temperatures in England were so low that 1816 was called "the year without a summer." London temperatures were 5°F cooler than normal. Meanwhile, New England reported snow in June, frosts every month, and food shortages because crops did not ripen.

We can debate whether human activities alter global weather, but we know volcanoes do. Some eruptions have cooled the earth by almost 1°F for a full three years. Volcanoes emit large quantities of carbon dioxide, increasing atmospheric greenhouse gases. At times, they release enough particulates and sulfur dioxide gas (which transforms into sulfate aerosols) to overwhelm industrial contributions. Volcanoes also emit huge masses of chlorine, and the halogen gases that take the blame for destroying ozone.

There is no doubt that volcanoes are villains when it comes to climate modification, but we remain unsure of their total impact. Volcanic measurements are difficult to execute, and even more difficult to pay for. Scientists who want to measure smokestack outputs in places like Pittsburgh may get money from the EPA. But the scientist who wants to watch a volcano erupt in Java sounds like someone trying to get paid for an exotic vacation. Besides, there is no agency in charge of controlling or assessing volcanoes.

Volcanoes are highly individualistic. Some go off with a bang every time; others growl a while before erupting into a roar. The lava from one differs from the lava in the next. Some put out large amounts of gases, others do not. These differences mean that estimates of volcanic effects on climate are highly approximate. In fact, unless the output of each volcano is measured (there are over 500 active volcanoes with perhaps 20 erupting in any given year), the estimates are not even estimates—they are guesses.

Scientists now have techniques for estimating what gases spewed forth during eruptions that occurred long ago. One technique involves collecting the small glass spheres that formed from molten rock during the eruptions. Researchers measure the gases stored in the spheres, and relate it back to the volcanoes' output. Such results are subject to large errors, but large errors are better than wild guesses. Even with measurements of present-day volcanoes, the concentration of carbon dioxide released is hard to measure. Scientists have to use ratios of carbon dioxide and sulfur dioxide to deduce how much carbon dioxide an eruption produces.

World temperatures decrease when volcanic ash enters the stratosphere. The small particles stay in the stratosphere for a long time because rain, which originates at lower altitudes, cannot cleanse them from the air. The particles may take part in ozone destruction. Volcanoes also release chlorine and bromine into the stratosphere, and both are significant ozone destroyers. Aerosols from volcanoes must be shot into the stratosphere if they are to affect global weather, but the greenhouse gases released are important even if they never reach the stratosphere. Lava, on the other hand, makes spectacular television images but has no known global climatic effects.

The lava and ash released by a volcano are the easiest quantities to measure. Lava and ash volumes are also the primary measurements available for eruptions that occurred in the past, because lava stays around. Past estimates are important because volcanoes may wait hundreds if not thousands of years between eruptions.

Some volcanoes put out large quantities of lava, but blow nothing high enough into the atmosphere to block sunlight. Some volcanoes release a larger fraction of polluting gases than others. Almost all volcanoes release a mixture of gases, but even the mixture from a single volcano changes as the eruption proceeds. The first puff is often quite different from later emissions, especially when the volcano goes off with a bang. The measurement problem, then, is to learn just what comes out of each volcano and what fraction of it gets into the stratosphere. The fraction is particularly difficult to determine, because most of the material that reaches the stratosphere is released soon after the eruption starts. In fact, most of the stratospheric mixture may be released before anyone realizes the eruption is taking place, and before scientists arrive on the scene. At Mt. St. Helens, particles had almost stopped reaching the stratosphere within half an hour after its big blast.

VOLCANIC DUST
AND THE WEATHER

A great cleansing mechanism operates in the lower atmosphere: rain. Raindrops capture dust particles and carry them back to the ground. Chemical balances within the overall atmosphere may take a long time to recover from an insult, but the lower atmosphere cleans particles out within days or weeks. The predicted disaster resulting from the Kuwait oil well fires did not materialize because the smoke stayed in the lower atmosphere, and the lower atmosphere took care of itself.

In the stratosphere, just above the tropopause, aerosols are another story. No rain falls at these altitudes. Small particles can float there for years. The key word here is "small." Larger particles fall on their own. Figure 1 indicates how quickly particles of different sizes fall. From the graph, it appears that micrometer-sized particles could stay in the stratosphere virtually forever. They don't, however, because the particles grow. Some bump into other particles and stick together. Others grow because sulfates and gases condense out of the air onto the particles' edges, slowly increasing their size.

Both volcanoes and coal-burning

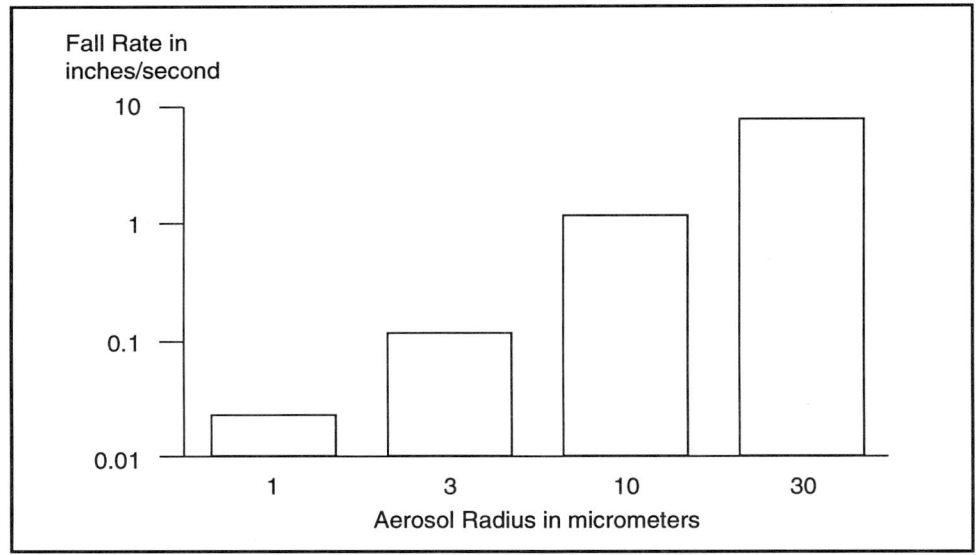

FIGURE 1   *Small dust and aerosol particles fall quite slowly. This means that small particles stay in the air until rain sweeps them away. Since no rain falls through the stratosphere, volcanic ash and other particles can remain up there for years.*

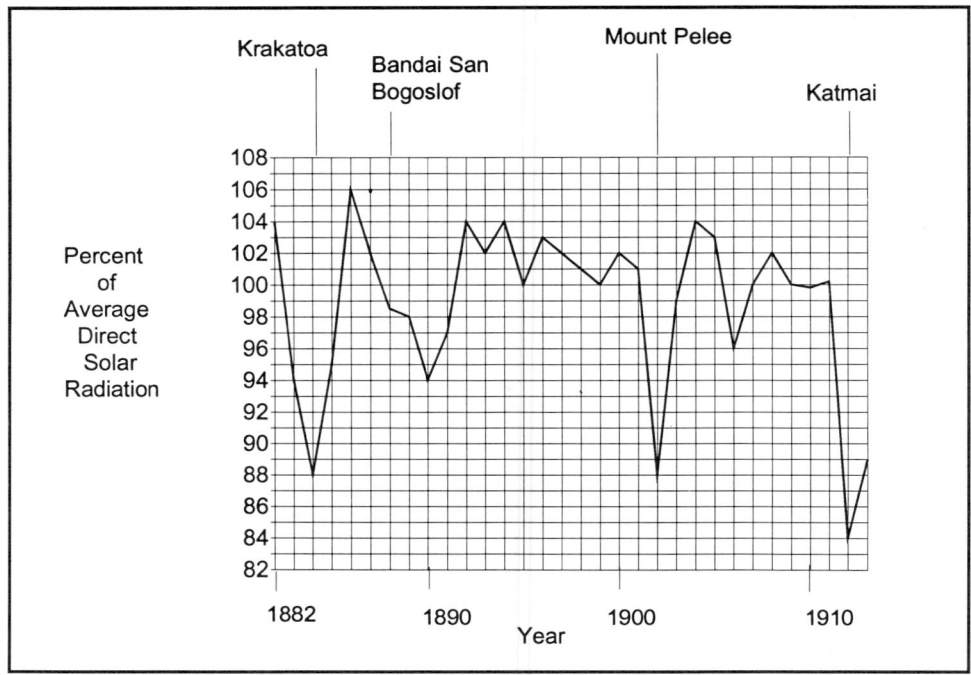

FIGURE 2   *Even at the turn of the century, figures such as this appeared, showing how volcanic eruptions caused decreases in sunlight at the ground. Volcanic ash in the stratosphere scatters some sunlight back into space and decreases heating at the ground.*

power plants emit sulfur dioxide, a gas that causes acid rain. After release, the gas quickly condenses into particles. Experience suggests that sulfates shot into the upper atmosphere by volcanoes will last for about three years. These sulfates are largely responsible for the cooling effect of volcanoes.

The idea that volcanoes block the sun and create cooler weather is not a new one. Curves like Figure 2 were being published at the turn of the century. The figure shows the decrease in direct solar radiation following volcanic events. The figure is correct, but it may make volcanic effects seem more important than they really are.

Here is why the curve is somewhat misleading: Direct solar radiation is the energy received in the direct solar beam. Pyroheliometers measure direct solar radi-

ation by looking only at a small portion of the sky, centering on the sun itself. As a result, their measurements exclude most scattered radiation. Scattered radiation in the atmosphere is what makes the sky blue, red, or whatever color it happens to be. Even when skies are clear, indirect or scattered radiation makes up ten percent of the total solar energy received. If the number of aerosol particles in the atmosphere increases, more radiation is scattered, because aerosols scatter more radiation than they absorb. Figure 2 does not show the decrease in total solar radiation; it shows only the decrease in direct solar radiation. About half this decrease probably still reached the surface as scattered or indirect radiation.

The graph also punches up its message by magnifying the percentage scale to make the changes more obvious. If the percentage scale started at zero, as most graphs do, the effect of volcanic scattering would appear much less impressive.

You might expect aerosols and smoke injected into the stratosphere to remain pretty localized, but this is not the case. Brisk winds like the jet stream blow constantly in the upper atmosphere. In about three weeks, a volcanic cloud can spread around the world. Winds diffuse the smoke because different altitudes have different winds. At some altitudes, the smoke is pushed faster and in different directions than at others. The winds move primarily in an east-west direction, so the north-south spread of volcanic ash takes longer. As a result, volcanoes erupting in the tropics are more effective at scattering sunlight and reducing incoming solar radiation. If a volcano erupts near the equator, its smoke can spread into both hemispheres. The lack of significant north-south winds across the equator keeps eruptions elsewhere from having much effect on the opposite hemisphere.

An amazingly small amount of material can cover the world with a layer of aerosol particles. Only about a million tons a year of the right particles might cool the world by 10°F. If and when the people of the world decide that things are getting too hot, we could cool the weather off fairly cheaply. Airliners could serve the dual purpose of carrying passengers and putting a veil of aerosols into the stratosphere. The jet engines could be adjusted to burn fuel inefficiently and produce small carbon aerosols. It might sound strange to suggest using oil to produce aerosols, but oil is among the cheapest of chemicals. What other chemical costs less than $20 per barrel? Carbon aerosols made from oil are highly effective at scattering light, yet chemically inert and unlikely to interact with stratospheric gases. Small particles can scatter the sunlight that warms the surface, yet remain too small to have any effect on the outgoing infrared.

Perhaps the most difficult aspect of using this method to turn down the earth's thermostat would be arriving at an agreement about how much temperatures should be lowered. Some nations would demand no cooling, some a little, others a lot. If one country decided to carry out a plan independently, it would undoubtedly find itself the subject of an embargo for messing up the weather. The countries of the world could never agree on how worldwide weather should be changed. As the 1992 environmental summit conference in Rio demonstrated, nations cannot even formally agree *not* to change the earth's climate.

Obviously, volcanoes do not ask whether we want cooling. They simply erupt when they please. Mt. Penatubo's eruption wiped out most of the temperature increase that had accumulated over the last 50 years—the same increase that served to demonstrate that global warming was underway.

In 1970, a British meteorologist named Lamb undertook a detailed study of volcanic materials that had entered the stratosphere over the last few hundred years. His results are shown in Figure 3. Along the right side of the figure is an arrow showing the amount of material required for a 1°F decrease in global temperatures. The values of loading in the figure and the amount mentioned above for cooling the world by releasing aerosols are different. The loading shown in Figure 3 is actual loading by volcanoes. By optimally selecting aerosol sizes, smaller quantities of material could provide more cooling than the amount provided by volcanic materials.

According to Lamb's computations, volcanic materials have a greater effect

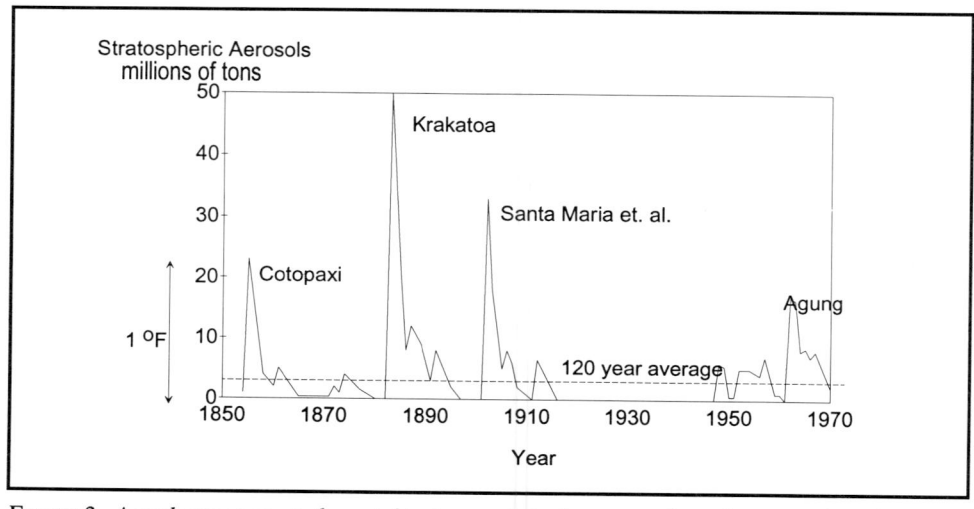

FIGURE 3   *As volcanoes erupt, the weight of aerosols in the stratosphere shoots up, then slowly decreases. The labels give the names of volcanic eruptions corresponding to the peaks. Along the left is an arrow showing the amount of aerosols needed to change ground-level temperatures by 1°F. Notice that there were few eruptions in the first half of the 20th century.*

than all the aerosols introduced into the stratosphere through human activities. This is a different number, though, than the total mass of aerosols we have produced or the amount of heating we have caused through increased greenhouse gases. Values in the figure concern only the stratosphere, and most of our aerosols never reach as high as the stratosphere.

Note that the amount of volcanic cooling has been much less since 1910 than it was between 1850 and 1910. Lamb carried out his study in 1970, so more recent volcanic events are not included. Even after considering the activity since 1970, one must conclude that volcanoes affected weather to a lesser degree in this century than in the 19th century. There is no reason to think volcanic activity is decreasing. It is just random by nature, and the recent past represents a lull period. Nor can we predict an increase for the near future. Random processes are quite simply that—unpredictable.

Penatubo's recent eruption shot a rich supply of particles into the stratosphere. Even though the resultant cooling was more than enough to wipe out all temperature increases accumulated during this century, this is a short-term effect. In three or four years, the sulfates from Pinatubo will be gone. Temperatures may rise again, unless another volcano releases more aerosols. Penatubo's eruption changed stratospheric circulations, and that caused a nine percent decrease in ozone. We believe that the ozone will come back when the aerosols are gone, but we can't be certain. No one thought changes in circulations would occur; no one anticipated the large decrease in ozone.

Not all major eruptions shoot aerosols into the atmosphere. Mt. St. Helens, for example, erupted with a powerful explosion and produced a large volume of ash, but worldwide temperatures remained unchanged. One reason may have been the geographical location of Mt. St. Helens. It is too far north to send aerosols into the equatorial region, which is so important in the earth's heat budget. Measurements there also indicated that the eruption injected a relatively small quantity of material into the stratosphere. Much of the stratospheric material consisted of large particles that fell rapidly enough to settle quickly out of the upper atmosphere.

GASES FROM VOLCANOES

We often visualize the lava and smoke that spew out of volcanoes, but we rarely consider the gases that emerge, unless we happen to hear about a release of poisonous volcanic gas. The fact is, volcanoes release a great deal of gas. If not for the gases, volcanoes would not produce explosions. When Mt. St. Helens exploded and blew hundreds of feet off the top of the peak, the explosion was a result of gaseous emissions. Likewise, gasless volcanoes would never spew ash into the stratosphere. Hot, rising gases actually carry the ash into the upper atmosphere. How hot the gases are determines how high the ash rises.

Many people think of lava and magma as two names for the same thing, but this is incorrect. Lava is the material that spills over and runs down the mountain side. Magma is material that emerges from deep within the earth. A separation takes place that produces lava from magma. After magma reaches the surface, its gases spew out, perhaps carrying ash with them. Whatever is left behind is called lava. Many volcanoes produce pumice and volcanic rock so full of gas that it can float on water. Both pumice and floating rocks occur because the gases puff up the magma. While still underground, the material exists as either highly compressed gas or in a non-gaseous phase. Non-gaseous phases might include liquefied gases and dissolved gases. Under high pressure, some gases remain dissolved in the molten magma. Rising to the surface releases the pressure, and the gas expands.

It is possible that most of our atmosphere was created by volcanoes. Whether the world had an atmosphere when it was new is one more thing about which scientists cannot agree, but the absence of neon in the air argues against the idea. Neon, like helium, is a noble gas that forms few or no chemical compounds. Neon does not condense at atmospheric temperatures or escape from the atmosphere. Helium is light enough to drift right out of the atmosphere into outer space, but neon is too heavy to do so. In other words, all the neon that was in the atmosphere when the world came into being should still be there. In fact, there is virtually no neon, so there was probably never any more than exists now. By contrast, a great deal of neon exists in the rest of the universe. If the world came equipped with an atmosphere, that atmosphere should contain whatever the universe contains, so the inference is that the earth came without an atmosphere.

The scarcity of atmospheric neon suggests that the atmosphere formed as the earth condensed. The upper crust of the earth cooled first, forming a solid layer of igneous rocks. The gases given off as deeper rocks solidified had to have a means of escape, and they did so primarily through volcanoes and hot springs. Materials released in this manner included the atmospheric gases and water. When the world was cooling, it was too hot for liquid water to exist. If the earth had no initial atmosphere, all the water in sedimentary rocks, the oceans, and the atmosphere had to come from somewhere inside the world.

Saying that volcanoes released enough gas to form the atmosphere seems plausible, but the notion that volcanoes also produced enough water vapor to rain and fill the oceans sounds much less likely. There was no liquid water on the surface of our red-hot planet. On the other hand, keeping oceans of water suspended in the atmosphere of a hot earth (if the earth had an atmosphere from the beginning) would present quite a problem, too.

Present-day igneous rocks contain about one percent water. To explain all the water in oceans only requires that the material that formed those rocks contained four percent water before they cooled and crystallized. The three percent water they lost in the course of solidifying would have drained away slowly, because the rocks cooled slowly, and that water had to find its way to the surface through some form of volcanic activity. Remember, the top cooled first and formed an impervious crust. As time passed, the cooling extended to deeper levels, causing the crust to thicken as new rocks solidified farther beneath the surface. Water excluded during solidification would appear deep under the crust. Something had to poke a hole in the solid layer of crust before the water could be released to form oceans, clouds, and water vapor in the air. Volcanoes were that something.

A strong argument for the theory that the atmosphere came from within, via volcanic action, appears in the pairs of bars in Figure 4. The left member of each pair represents the combined weight of a substance in the air, water, and sedimentary rocks of the earth. The right bar is the amount of the substance volcanoes presently produce every year, multiplied by the earth's age. In other words, the right bars represent how much material volcanoes may have spewed forth during the history of the world—assuming that volcanoes now spew at the same rate they always have, and that volcanoes are neither more common nor less common than they were in the past. This assumption is not as faulty as you might think. Volcanoes spewed these materials out, but

not all of them reside in the atmosphere. Most of the water exists in the ocean; much of the carbon lies in coal, oil, and limestone deposits. The atmosphere contains only that fraction of materials that the world has not yet placed into longer-term storage. Mother Nature does try to pick up after herself, but it takes her a while to put things back into the earth.

Using a logarithmic scale for the graph is a device scientists sometimes use to make the data appear to agree more strongly than it really does. Logarithmic scales make the one to 10 distance the same as the 10 to 100 and 100 to 1,000 distances. Figure 4 employs this device, but the agreement nonetheless seems too strong to explain through chance. Volcanoes can only account for half the water and chlorine, but that's not too far off considering how approximately we estimate the output of volcanoes. Notice that the substances listed do not include oxygen. Oxygen did not appear as an atmospheric gas until plants started generating it. At first, the atmosphere contained no uncombined oxygen. There was no ozone, either, since ozone is nothing more than three oxygen atoms bonded together. Therefore, life on earth evolved without an ozone layer to protect it from ultraviolet radiation.

Here is another caution about the data in Figure 4. Not all of the material emerges only during eruptions. Volcanoes emit gases even when they are not erupting. Previously deposited lava outgases and hot springs pull large quantities of volcanic gases and chemicals to the surface. Eruptions are spectacular, but volcanoes cook gases for hundreds of years between eruptions. Some magma that never reaches the surface interacts with subsurface water, heats it, and causes it to transport volcanic materials to the surface. Recent measurements around Mt. Etna in Sicily discovered that it, together with the lava it spilled in the past, put 25 million tons of carbon dioxide into the atmosphere each year, even when no eruption occurred.

The quantities in Figure 4 might evidence a stronger match if we knew more about submarine volcanoes—undersea volcanoes whose eruptions may never reach the surface. Many of their eruptions are unnoticed and unrecorded. Most of these volcanoes are undetected, so their output is difficult to estimate. Their emissions may not directly reach the atmosphere, but they eventually become part of the inventory of chemicals in the atmosphere, and in the sedimentary rocks formed in the oceans.

Two University of Michigan scientists, Robert M. Owen and David K. Rae, claim that hot water venting through the sea floor might have increased enough to trigger a greenhouse hot spell some 50 million years ago. Owen and Rae also suggest that the vents resulting from undersea magma now account for a fifth of the world's naturally produced carbon dioxide. According to their calculations, rearrangement of the earth's tectonic plates might have quadrupled the amount of calcium emitted from the vents. That calcium increase might have interacted with sea water, causing atmospheric carbon dioxide to double.

If we could conclude that volcanoes release more carbon dioxide than man, then we could continue to burn fossil fuel with impunity, convinced that our small additional contribution would be unimportant. However, by all systems of calculation, the emissions of carbon dioxide from volcanoes represent no more than a small fraction of the anthropogenic releases. Volcanoes release about one percent as much carbon dioxide as humans. Man and plants continually recycle carbon that volcanoes released long ago. Volcanoes release the carbon only once, but nature can transform it many times.

Fossil fuels come from carbon dioxide spewed out by volcanoes in the past. Volcanoes emitted carbon dioxide into the

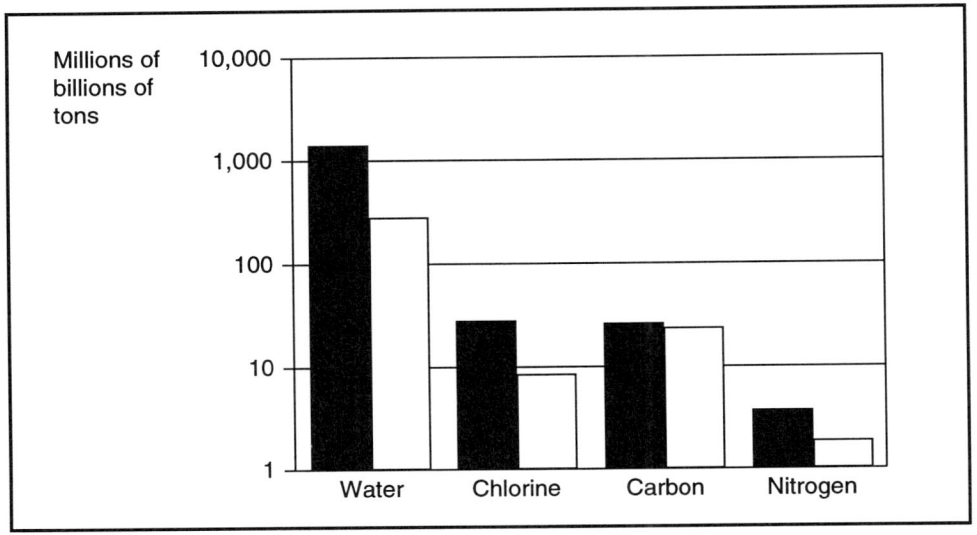

FIGURE 4 *The dark bars show the total weight of different substances in air, oceans, and sedimentary rock. The white bars show the total amount produced by volcanoes since the earth's crust formed. This assumes that volcanoes have always spewed out materials at the same rate as they have in recent history.*

atmosphere. Plants used that carbon dioxide in order to grow, and this growth headed the food chain for animals. Remains of plants and animals, containing carbon, were deposited in the earth, where pressure, temperature, and time transformed them into oil and coal. Some carbon used in the formation of fossil fuel came from volcanoes. Much of it probably came from the erosion of rocks. Geochemists can and do debate the fractions of each.

VOLCANOES AND OZONE

The effect of volcanoes on ozone is seldom discussed, despite the fact that volcanoes could greatly affect ozone concentrations. As Figure 4 indicates, volcanoes release large quantities of chlorine, and chlorine is the chemical that causes the ozone hole. Solid particles, such as those that volcanoes inject into the stratosphere, seem to be necessary before chlorine becomes a truly effective destroyer of ozone in the stratosphere. Release of solid particles into the stratosphere is a specialty of volcanoes—so are volcanoes and ozone one more thing we should worry about?

The eruption of Mt. Pinatubo relieved a major concern about volcanic eruptions and the ozone hole. Remember that the ozone hole forms over the Antarctic region because that is where ice particles form at altitudes of 50 miles. Only there is the upper atmosphere cold enough to produce ice particles from even the slight amount of water that remains at such high altitudes. Ice particles seem to accelerate the destruction of ozone performed by chlorine.

Until Pinatubo erupted, scientists worried whether ice particles were unique, or if any sort of particle might serve as the platform for ozone destruction. If other kinds of particles could assist in the ozone destruction, the situation was more serious. High-altitude ice particles can only

occur over the Antarctic (and in the Arctic, but scientists cannot agree on why there is no Arctic ozone hole in spite of high ice clouds forming there). Volcanoes, however, form their own clouds by injecting ash into tropical and mid-latitude regions. Thus, a big supply of volcanic particles in non-polar regions could destroy ozone outside the Antarctic— throughout the year and over a wide range of altitudes. The ozone hole might then spread throughout the atmosphere, which would mean disaster.

By injecting sufficient particles into the stratosphere, Pinatubo provided a test scenario. We know now that the ozone has endured. It appears that the particles may have decreased mid-latitude ozone by a small amount, but no massive decreases occurred. The threat, which never received much publicity, was apparently not a threat at all.

Chlorine is difficult to deal with. Those who claim that there is no current threat to the ozone point to the large quantities of chlorine released by volcanoes. They argue that the ozone has co-existed with this chlorine for eons. Thus, they say, chlorine does not represent a threat to ozone.

Dixy Lee Ray maintains that volcanoes dispel the notion that ozone is destroyed by CFCs. She is an admirable person— among other things, she is a woman who made it to the top of the scientific world before we became concerned with women's rights. Ray started as a zoologist. She was accomplished enough to head the Atomic Energy Commission (AEC) back in the days when the AEC supervised the development of nuclear weapons and nuclear reactors. Later, she entered politics and became governor of the state of Washington. She then returned to science.

In her book entitled *Trashing the Planet*, she points out that, in 1976, Mt. St. Augustine in Alaska injected over 100 billion pounds of hydrochloric acid (nearly all the weight is due to the chlorine in the acid) into the atmosphere. This amounted to 570 times the 1975 world production of CFCs. Ray also noted that Mt. Erebus, just upwind of the Antarctic's McMurdo Sound, has been erupting continuously for the last 100 years, releasing 1,000 tons of chlorine every day. The fact that Mt. Erebus is situated in the Antarctic gives Ray's argument special force.

Chlorine is among the most reactive of all the elements, which is why CFCs are interesting. If you release any normal chlorine compound into the atmosphere, the high reactivity of the chlorine will ensure that the compound disappears from the atmosphere with little delay. Before the chlorine compound reaches the stratosphere, it will generally react with something in the air and return to the earth.

CFCs violate this rule, because they interact with nothing in the atmosphere. CFCs eventually find themselves diffused into the stratosphere, where solar ultraviolet radiation finally provides the power needed to tear the CFC molecules apart. After the sun tears the molecules apart, the chlorine is no longer inactivated, so it is free to destroy ozone.

Volcanoes emit simple inorganic chlorine molecules, not CFCs. The chlorine and bromine emitted by volcanoes are highly reactive. Members of the chlorine family of chemicals released into the lower atmosphere by volcanoes react with substances and drop back to earth. However, that is not the full story.

What about the chlorine and bromine sent directly into the stratosphere during violent phases of eruptions? Only a fraction of total emissions reach the stratosphere, but that small quantity may destroy some ozone. Perhaps such destruction even keeps the world in equilibrium—we might have too much ozone without the volcanoes. Still, measurements of chlorine in the upper atmosphere show that concentrations are increasing rapidly, and the number of volcanic eruptions does not explain the rapid increase. We are left with human activity (producing CFCs) as the best explanation for the increases in stratospheric chlorine. We are left, also, with responsibility for the ozone hole.

Now back to Mt. Erebus. In winter, the sunlight that heats the surface to cause convection and create the tropopause disappears from the Antarctic, thereby lowering the altitude of the stratosphere. Since Mt. Erebus is located in the Antarctic, a high proportion of its emissions could go directly into the stratosphere. This might contradict the argument that volcanic chlorine rarely or never reaches the stratosphere.

Do volcanoes prove that CFCs are not responsible for the ozone hole? Probably not. Most scientists believe that CFCs are the culprit. Although the evidence pointing to CFCs is largely circumstantial, the circumstances are strong. Considering the risk posed by loss of ozone, it seems wise not to gamble that CFCs are falsely accused. Loss of ozone is not a criminal case requiring proof beyond a reasonable doubt, and we may be better off if we assume that we are all "guilty as charged."

VOLCANIC HEAT

Volcanoes release a great deal of heat, but that heat adds an insignificant amount to global warming. Heat from deep within the earth could be an important part of global warming, but in an offhand way. It could be important in reducing, rather than increasing, global warming if we began to use geothermal energy instead of fossil fuel to supply our energy needs. Geothermal heat could provide energy without releasing greenhouse gases. Geothermal energy means less greenhouse gases are released by power plants, and the reduction in greenhouse gas releases could reduce global warming.

Since the world became energy-conscious during the oil shortages of the 1970s, many people have wondered why we do not use flowing lava as a source of energy. Big volcanic eruptions may lay down a few cubic kilometers of lava (one cubic kilometer is 1.3 billion cubic yards), and that quantity represents a great deal of energy.

Lava temperatures are about 2,000°F. For some reason, lava never gets much hotter. If it gets much cooler, the lava solidifies. After a lava lake cools and looks solid, it is like a frozen lake—only a thin layer on top is solid. The outer crust is a good insulator that keeps the center molten for years. Twenty years after Kilauea's eruption, core samples taken from the lava lake formed during the eruption contained liquid lava at a depth of 150 feet. Even the surface crust stays hot for years. According to some stories, a man could light his cigar in the fissures of Mexico's Jorullo Volcano 20 years after its eruption in 1759. The stories don't specify how many burned noses accompanied these cigar-lighting experiments.

According to estimates, one cubic kilometer of lava could provide a city the size of San Francisco with electric power for 200 years. It sounds promising, but until now, use of lava as an energy source has only been discussed. One often-cited problem with implementation is the difficulty of building heat exchangers that can extract heat from the lava. This is puzzling. We can build nuclear submarines that use liquid metal heat exchangers to extract energy from their nuclear reactors, but we seem unable to build heat exchangers for the extraction of energy from lava.

In truth, there has been little interest in extracting heat from lava. Most lava flows are situated far from where the power is needed, and the flows exist in layers of different thickness. The lava could provide electric power only in particular locations. These limitations dampen the enthusiasm of power companies. If utility companies really wanted to use lava energy, heat exchangers could be developed.

Geothermal steam and hot water offer a more positive story. They have been and are being used in commercial operations. Iceland receives most of its home space heating requirements from geothermal hot water. Wells tap the hot water, pipes carry it throughout cities, and meters record how much is used by each household. Although cheap, the hot water is not free.

Electricity has been generated in the Geysers area north of San Francisco for years. In spite of its name, there are no geysers in the Geysers area—only steam vents. The location was first developed as a health spa, and the first electrical power generation in the area took place during the 1920s, when wells were drilled to produce steam at a pressure of 60 psi. Unfortunately, the commercial undertaking was a failure—not because of the steam, but because there was no demand for the electricity being generated. It was an idea ahead of its time.

In the late 1950s, large-scale development of steam wells began. The Geysers area now produces thousands of megawatts of electric power, while remaining a popular camping and recreational site. Most of the steam from the vents is captured instead of released, but occasional releases occur. When that happens, campers complain loudly about the foul odors that are a natural part of geothermal steam. Ironically, the hot springs producing those foul odors are what initially made the area popular for recreation years ago.

Commercial development of geothermal steam and hot water is now underway in the volcanic regions of Central America and elsewhere in the world. In some cases, the water is not hot enough to produce steam, but even semi-hot water can provide energy for power by putting the water through a heat exchanger. In the heat exchanger, hot water vaporizes some chemical, such as the CFC freon, and the vaporized chemical then drives a turbine for generating electricity. The chemical is recondensed and recycled, allowing it to make trip after trip through the heat exchanger. Our ability to use hot water instead of steam greatly expands the number of areas we can use to generate electric power. Even when the water is hot enough to provide steam, heat exchanges are sometimes necessary, because the steam contains so many minerals. If used directly, the steam would clog the pipes and precipitate onto the turbine blades.

Geothermal heat comes from the magma itself, not from lava. Apparently, magma may at times force its way up to within five or ten miles of the surface. The magma is too deep and probably too hot for direct utilization, but it releases heat that raises the temperature of the rocks and water directly above. The rocks and water can then be tapped for heat.

By far the greatest number of the world's geothermal regions contain hot rocks but no water. Los Alamos National Laboratory has devoted several years to developing the technology for extracting energy from such waterless hot rocks. The idea is to develop ways of working with hot, dry rocks, thereby expanding the number of viable regions and the amount of energy available. The technique developed at Los Alamos sounds simple enough. Two wells are drilled into the hot, dry rock. Pumps send cold water down one hole, and hot water comes up the other. The two wells are situated a few hundred feet apart. The hot rock is nonporous, so

water does not leak away, but it does not flow from well to well, either. After the wells are drilled, high-pressure pumps fracture the rock. Oil companies use the same fracturing technique to increase the flow from oil and gas wells. Fractures must spread out from the first well and intercept the second well before water can flow between the two wells. Water pumped down the first well is heated by the rocks as it flows along the fracture to the second well.

A great deal of heat can be obtained from the water flowing through this crack, but the rocks around the crack will eventually cool. If the technique is to be fully successful, the cooling effect of the water must cause additional cracks to develop, just as pouring cold water in hot drinking glasses often cracks the glasses. The cracks would then allow ever-larger volumes of rock to provide heat.

Energy from geothermal sources may not be pollution-free. Hot water or steam can contain many minerals. An extreme example involves water from the Salton Trough in California. Brine from this reservoir contains 25 percent dissolved solids. Sometimes, disposal of minerals merely requires pumping the minerals back into the wells from which they came, but that can be an expensive process. In spite of difficulties, Salton Trough water is now being used to generate electricity. Most of the chemicals in the water are collected and trucked to disposal areas.

The contrast between geothermal power and nuclear power is an interesting one. Many people oppose nuclear power, but in fact it is nuclear decay that keeps the center of the earth hot. Thus, geothermal energy is an indirect form of nuclear energy. For that matter, solar energy is nuclear energy, too. It is produced by the thermonuclear reactions that drive the sun.

The use of geothermal power produces

no radioactive wastes that must be buried for generations but, as is the case with nuclear reactors, geothermal energy comes from the splitting of nuclei. Presumably, some radon gas emerges from volcanic eruptions or geothermal steam and water, but radon gas leaks through the ground everywhere. It may be more concentrated in volcanic areas.

And so volcanoes, like so many of Mother Nature's creations, are part hero and part villain. They are powerful and unpredictable polluters, and they have been known to kill thousands of human beings in one fiery eruption; however, they are sites of great natural beauty, and they may one day provide us with cheap, relatively clean sources of geothermal energy.

▲ 12 ▼

Putting It All Together

Can we eat our own stew?

In the previous chapters, we covered the major forces that affect our climate. Some, like volcanoes, are not subject to change engineered by humans. Others, like cloud cover, we may already be changing. We can be certain that we are changing greenhouse gases and ozone.

All these forces act together to form our climate. We know this, but we do not know exactly how the forces fit together. Examining oceans, volcanoes, clouds, and other factors helps us appreciate the complex, unpredictable nature of our climate. In trying to assess things realistically, it is also useful to realize that scientists know less than we like to think they do. We may find ourselves longing for a candle in the dark, but no such candle can realistically be expected in the near future. We must instead choose our path with no more than a glimmer of information at our disposal.

The path to reasonable action is illumined by the knowledge that predictions of the future are often scary by design. If someone makes a boring prediction, no one pays attention. Still, frightening predictions are out there, and in many cases no one can prove or disprove them. We

are forced to wait and see what happens. Our choice is this: We can wait and do nothing, or we can try to do something positive as we wait.

One primary positive action involves cutting back on our emissions. We are cutting our CFC production, which proves that worldwide agreements are possible. Cutting out CFCs is easier than cutting back on carbon dioxide. Refrigerators continue to work for a time after CFC production is stopped, so we experience little or no immediate pain. Take away our central heating, air conditioning, or cars, and the pain—or at least the inconvenience—is immediate.

Future carbon dioxide releases in the United States may be headed for a cutback in any case. Incomes are drifting downward. With less income, Americans cannot afford to consume as much.

Few care to contemplate this, but a drastic reduction in our standard of living may be the only way we can hold carbon dioxide levels where they are. We could probably slow our increase enough to keep releases at current levels of increase. But overall carbon dioxide was increasing even at 1950 emission levels, and the atmospheric increase began much earlier. No one suggests that we can cut back to such low levels. The IPCC reports that cutting our emissions by 60 percent would stabilize carbon dioxide at its current levels.

Instead of reducing our standard of living, we could choose to redefine it. Right now, we define "standard of living" in terms of number of cars, size of houses, and other indicators of consumption. If we defined our standard of living as the purity of the air we breathe, the number of birds we hear singing, and the time we spend with our families, a completely different set of values might evolve. Right now, government figures imply that those who consume less are living a sub-standard life. And yet most of us would trade the big house in the city, the long work hours, the traffic, and the impersonal corporate lifestyle for a small cabin in the woods.

The carbon dioxide budget is much like the federal budget deficit in the U.S. It is easy to cut a small amount and claim that progress is being made, while continuing business as usual. As with the federal budget, some carbon dioxide releases are easily cut. However, implementing cuts that are deep enough to solve the real problem means real pain. Building new power plants and new public transportation systems, and modifying homes so that they are more efficient, requires a huge amount of money. Only so much money is available. Spending it on cutting carbon dioxide emissions means we must spend less on other goods and services. We are too deeply in debt to do everything at once. Reducing—or redefining—our standard of living is necessary. If we spend money on VCRs, there is ultimately no money for new transportation systems. If we refuse to cut the federal deficit, we will most certainly refuse to change our standard of living.

The United States, of course, is not the source of all carbon dioxide emissions. We emit more than our share, but a worrisome factor is that many people in other countries would like to live the way we do. If they achieve their goal, they will emit as much as we do. As the Third World countries try to improve life for their rapidly growing populations, their emissions will shoot up. Impoverished populations generally pay little attention to problems that may lie 100 years down the road. They strive to live a better life right now.

Perhaps this sounds terribly pessimistic. Pessimism and realism can be hard to separate. Looking at what people have done in the past and what they can be expected to do in the future makes it seem realistic to predict that no emission cutback will occur. However, there are more positive ways of regarding the situation. One positive outlook involves finding ways of living with temperature increases. After we endure the pain of significant temperature change, we can entertain our great-grandchildren with tales of how terrible the winters were way back when. Temperature increases and our adaptation to them might even allow us to expand production and feed more of our people.

BELIEVING PREDICTIONS

The world has always been full of predictors—soothsayers, fortune tellers, even prophets. Picking out which ones were prophets usually involved waiting until hindsight afforded us a clearer view. We are in the same situation today. There are many predictions, but no one knows which predictions are correct.

The predictions of global warming emerge from our best current scientific knowledge. They are probably wrong, but no one knows to what degree or in which direction. Things may turn out better or worse than the experts predict. Those experts who predict no change are almost certainly wrong, because nature never stays the same. Species and ice ages come and go. Deserts form, mountains erode. Humans may now be able to change the world's environment, but preventing it from changing on its own is well beyond our means.

The worst aspect of predictions is that they tend to evidence tunnel vision. Those who issue predictions seem always to miss the really important events. No one predicted the Irish potato famine. No one foresaw the coming of the AIDS virus. Malthus predicted doom from population growth without foreseeing how technology might increase the number of people we could feed. The little ice age must have represented a disaster, but it was not foreseen.

The problem with the predictions of global warming 100 years from now is that few of us can imagine life in 100 years. Even if the predictions are correct, none of us expect to be alive. Further, the prediction of higher temperatures a century from now must contend with other pessimistic predictions. Consider population levels, for example. If the current 1.7 percent annual increase in population continues, the world will contain five times as many people in 2092 as it does today. It is likely that such an increase would affect global lifestyles far more than a temperature increase of five degrees.

Extending the population figures even further reveals the limit of such predictions. At the present growth rate, the population would reach 157 times the current level in 300 years. Professor Albert Bartlett at the University of Colorado has observed that our 1.7 percent per year increase in population means that, in just 1,800 years, the weight of all the people in the world would equal the weight of the world itself. Obviously, the population will never become that large.

In the same essay, Professor Bartlett quotes estimates on oil reserves provided by the U.S. Geological Survey. According to these estimates, our present reserves, plus all the oil yet to be discovered, are sufficient to last just 60 years at 1984 rates of consumption. If that is true, because most of carbon dioxide increases result from oil, the present rate of CO_2 increases cannot continue for the 100 years often used in projections.

Unforeseen changes that might result from increased greenhouse gases are as worrisome as the predicted warming. The only defense we have against unexpected changes involves cutting back on use; even a small cutback may help. Who knows which straw will break the camel's back— or in this case, trigger an unwanted climatic change?

Here in the United States, it seems to me that we have been in a negative mood for most of the second half of this century. We see problems, instead of opportunities. It is time—in fact, it is *past* time—for a return to the positive outlook that made this country successful. If we assume a positive outlook, we take back the power of human resolve. We begin to speak differently to ourselves:

"Our infrastructure is old and falling apart," we might say. "Let us rebuild it in ways that reduce energy consumption and carbon dioxide emissions."

We profited by building this country. Why not profit once again by rebuilding and improving it? We must dream big— for our own benefit and for the benefit of others around the world. Is the sky really falling? We can't know for certain. In the end, it doesn't matter. Whatever the state of the sky, we must begin to make responsible choices here on earth.